SCIENTIFIC REALISM

NICHOLAS RESCHER

University of Pittsburgh

SCIENTIFIC REALISM

A Critical Reappraisal

D. REIDEL PUBLISHING COMPANY

A MEMBER OF THE KLUWER ACADEMIC PUBLISHERS GROUP

DORDRECHT / BOSTON / LANCASTER / TOKYO

Library of Congress Cataloging in Publication Data

Rescher, Nicholas.
    Scientific realism.

    (The University of Western Ontario series in philosophy of science ;
v. 40)
    Bibliography: p.
    Includes indexes.
    1. Science—Philosophy. 2. Realism. I. Title. II. Series.
Q175.R3943   1987   501   87-16419
ISBN 90-277-2442-3
ISBN 90-277-2528-4 (Pallas : pbk.)

Published by D. Reidel Publishing Company
P.O. Box 17, 3300 AA Dordrecht, Holland

Sold and distributed in the U.S.A. and Canada
by Kluwer Academic Publishers,
101 Philip Drive, Norwell, MA 02061, U.S.A.

In all other countries, sold and distributed
by Kluwer Academic Publishers Group,
P.O. Box 322, 3300 AH Dordrecht, Holland

*Also published in 1987 in hardbound edition by Reidel
in the series University of Western Ontario Series
in Philosophy of Science*

To
Robert Almeder

as a memento of many hours of
interesting and profitable discussion of these issues

TABLE OF CONTENTS

Contents

PREFACE

The increasingly lively controversy over scientific realism has become one of the principal themes of recent philosophy.[1] In watching this controversy unfold in the rather technical way currently in vogue, it has seemed to me that it would be useful to view these contemporary disputes against the background of such older epistemological issues as fallibilism, scepticism, relativism, and the traditional realism/idealism debate. This, then, is the object of the present book, which will reconsider the newer concerns about scientific realism in the context of these older philosophical themes.

Historically, realism concerns itself with the real existence of things that do not "meet the eye" – with *suprasensible* entities that lie beyond the reach of human perception. In medieval times, discussions about realism focused upon universals. Recognizing that there are physical objects such as cats and triangular objects and red tomatoes, the medievals debated whether such "abstract objects" as *cathood* and *triangularity* and *redness* also exist by way of having a reality independent of the concretely real things that exhibit them. Three fundamentally different positions were defended:

(1)   *Nominalism*. Abstracta have no independent existence as such: they only "exist" in and through the objects that exhibit them. Only particulars (individual substances) exist. Abstract "objects" are existents in name only, mere thought-fictions by whose means we address concrete particular things.

(2)   *Realism*. Abstracta have an independent existence as such. Abstract objects ("universals") are things every bit as real in their own way and manner as are cats and tomatoes. To be sure, the existence of such things transpires in a separate non-spatial, non-temporal, and non-material realm; their reality is not a *physical* reality.

(3)   *Conceptualism*. While abstracta do not exist independently and as such, they have a quasi-existence that does not wholly depend on the objects that exhibit them. This quasi-existence takes a *conceptual* form: they "exist" through their being conceived in the minds of people who are naturally disposed

to group together various items that roughly answer to these conceptions (those various creatures we group together as triangles or dogs, etc). Universals are thus mind-made all right, yet not arbitrarily, but under the guidance of certain natural dispositions of the mind in regard to the subject-matter at issue.

The medieval realists contended, the nominalists denied, and the conceptualists hedged on the idea that an adequate understanding of the world required us to accept the reality of such *abstracta* as properties, species, general, relations, and modalities.

Contemporary realism follows in the footsteps of these medieval deliberations. However, it shifts their focus from *universals* (the machinery for the descriptive characterization classification of natural things) to the *theory-creatures of natural science*. Contemporary realists are less concerned with unobservable *abstracta* – like shapes or quantities or colors – than with the unobservable "theoretical entities" of modern science like electrons or genes or magnetic fields. But after allowing for this change of focus, we still arrive at analogues to the three positions considered above:

(1)   *Instrumentalism*. The theoretical entities of natural science do not exist at all; they are merely useful thought-fictions that we invoke in providing explanations for observable phenomena.

(2)   (*Scientific*) *Realism*. The theoretical entities of natural science do actually exist essentially as scientific theorizing characterizes them. They are real items of the world's furniture and do indeed possess substantially the descriptive constitution ascribed to them by science.

(3)   *Approximationism*. While the theoretical entities envisioned by natural science do not actually exist in the way current science claims them to be, science does (increasingly) have "the right general idea". Something roughly like those putative theoretical entities does exist – something which our scientific conception only enables us to "see" inaccurately and roughly. Our scientific conceptions aim at what exists in the world but only hit it imperfectly and "well off the mark". The fit between our scientific ideas and reality itself is loose and well short of accurate representation. But there indeed is some sort of rough consonance.

These alternatives span the range within which the contemporary debate about scientific realism proceeds.

The present book argues *both* against an anti-realism of the instrumentalist variety *and* against a scientific realism of the sort just indicated. Instead, it defends a position within the orbit of the third approximationist view. The realism it espouses is one of intent rather than achievement – a realism that views science not as actually *describing* reality but as merely *estimating* its character.

Virtually every chapter of the book has been presented as a public lecture on some occasion or other over the period of its initial preparation during the 1984–1985 academic year. I am grateful to those who have participated in the ensuing discussions for helping me to work out the ideas more clearly. Moreover, I am indebted to James Allis for reading the book in its earliest manuscript version and offering helpful suggestions towards its improvement. And I am grateful to Christina Masucci, who competently and patiently saw the matérial through numerous revisions on the word processor.

Pittsburgh
October 1986

CHAPTER ONE

PROBLEMS OF SCIENTIFIC REALISM

SYNOPSIS. (1) Scientific realism maintains that natural science provides descriptive, accurate information about physical reality – that the objects of science exist as science claims them to be. (2) This position presupposes the essential correctness of natural science as we have it. But when science is seen in historical perspective, it becomes clear that there is no adequate justification for thinking that natural science as we now have it is actually correct. (3) Nor does it seem warranted to suppose that a *future* juncture will be reached when the science of the day correctly characterizes physical reality.

1. SCIENTIFIC REALISM

Brand Blanshard tells the story of how his Oxford tutor H. H. Joachim once asked him while on an afternoon walk, "Do you suppose that there really *are* such things as atoms?". Prior to World War I, this may have seemed a genuinely problematic issue. But in due course, after Hiroshima especially, it ceased to seem plausible to question the existence of atomic particles. To all appearances, the progress of science and technology has transformed the situation.

Philosophical realism as a general doctrine maintains the thesis that there is a domain of mind-independent existence and that we can obtain *some* reliable knowledge of it. But where are we to turn for such reality-characterizing knowledge? The most attractive and plausible line of response is that *natural science* is our best route to information about objective reality. This, at any rate, is the pivotal idea of the doctrine of scientific realism. And so, the exponents of this position hold that natural science affords accurate and reliable information about reality. If we want to know about the kinds of things there are in the world and the sorts of properties they have, it is to science that we should turn.

Such a position moves well beyond a generalized metaphysical realism that goes no further than maintaining that there is a mind-independent reality and that we can know *something* about it. For scientific realism moves on to say (i) that we can come to know a great deal about it; (ii) that this knowledge relates not just to peripheral matters but to essentials; and (iii) that this information is provided through science. Science gives an appropriate account of the salient and

1

characteristic features of what objectively exists in the real (mind-independent) world. The theories of science regarding non-observable entities – sub-atomic particles, electromagnetic fields, gravitational space-warps, and the rest – characterize the actual properties of real things in the real world, things every bit as real as the animals and plants and rocks that we see "with our own eyes".

Some writers on scientific realism make the issue pivot on reference rather than truth – on the applicability of the terms that figure in scientific theories rather than the acceptability of the theories themselves.[1] On this approach, the pivotal question is not whether one can make true statements about electrons ("Electrons carry a negative charge"), but whether that term at issue, "electron", actually designates – whether it bears referential weight in that there indeed are electrons to which this term applies. However, in our present context – that of *physical* realism – this distinction is ultimately one without a difference. The truth of abstract mathematical statements or of discourse about abstract principles, such as the rules of chess, does not preempt the issue of the *real existence* of such objects (i.e. numbers and rules). But where physical things are at issue, the case is different. In the case of a theory about physical object thing-kinds (electrons, black holes, magnetic fields), the "truth of the theory" and "the existence of its objects *as described* by the theory" are simply different sides of the same coin. To accept a scientific theory about little green men on Mars is *ipso facto* to accept little green men on Mars – and (in the circumstances) conversely. We cannot, in the case of natural science, be realists about theories (holding that, where true, they are true about reality) and yet fail to be realists about the *objects* at issue in the theories.[2] From the standpoint of realism, scientific theories and their objects stand or fall together.

2. THE PROBLEMATIC CHARACTER OF SCIENTIFIC REALISM: CURRENT SCIENCE DOES NOT DO THE JOB

The analogy between the medieval realism of universals and this modern realism of theoretical entities is instructive. Both forms of realism rest on a common basis. The existence of *some* things is evident: pine trees, for example, and domestic cats – to say nothing of ourselves. The existence of other things is not evident but rather inferential. We secure knowledge of these latter from the former via the principle:

(P)   Whatever is needed to provide an adequate account for the existence or the nature of something real is itself real and, as such, actually exists.

It is by this common principle that we moderns validate the claim that electrons exist, and that the medievals validated the claim that universals exist.

The difference between the two lies in the varying construction placed on the pivotal idea of "accounting for". The medievals took this to mean: "needed to 'account for' in the order of understanding – of *hermeneutic* explanation". The moderns construe it to mean "needed to 'account for' in the order of causality – of *causal* explanation". The medievals reasoned:

– Sundry types of particulars are obviously real.
– Universals are needed to provide an adequate explanatory account of these types of particulars.
– Therefore, by principle (P), universals are real.

The moderns reason:

– Sundry macro-objects are obviously real.
– Sub-atomic particles are needed to provide a causal account of macro-objects.
– Therefore, by principle (P), sub-atomic particles are real.

The medievals stood committed to a rationally intelligible world order. By contrast the moderns stand committed to a world-order that is causally explicable. Medieval realism and latter-day scientific realism are united by a common recourse to principle (P) to provide the linking premiss needed for the transition from patent to covert reality – from the palpably real to an order of suprasensible reality that is not evident at the crude level of ordinary sensory experience.

But this perspective highlights the inherent difficulty of scientific realism. For in order to apply principle (P) to establish the existence of non-observable entities one needs a premiss of the form:

– Non-observable entities of the type X *are indeed needed* to account for certain obviously real objects in causal terms.

However, we are never in a position to secure just this premiss since we
cannot glimpse future science to see what the materials of a definitive
account really are. The best we can ever do is to secure that non-
observable entities of the type *X are in the present state of science
thought to be* needed to account for certain obviously real objects in
causal terms. The italicized qualification is a concession to the realities
of our epistemic situation which cannot be eliminated. And this fact
blocks our ever using principle (*P*) straightforwardly and without establishing the reality of theoretical entities. This consideration renders
scientific realism problematic, seeing that it presumes our knowledge of
the essential correctness of natural science.

Scientific realism is the doctrine that *science describes the real world*:
that the world actually is as science takes it to be and that its furnishings
are as science envisions them to be.[3] Accordingly, scientific realism
maintains that such theoretical entities as quarks and electrons are
perfectly real components of nature's "real world". They are every bit
as real as acorns and grains of sand. The latter we observe with the
naked eye, the former we detect by complex theoretical triangulation.
But a scientific realism of theoretical entities maintains that this difference is incidental. In principle, these "unobservable" entities exist in
precisely the same way in which the scientific theories that project them
maintain. On such a realistic construction of scientific theorizing, the
declarations of science are factually true generalizations about the
actual behavior of real physical objects existing in nature.

But is this a tenable position? Clearly it has difficulties. For the
theoretical entities envisioned by current science will only exist *as
current science envisions them* insofar if current science is correct – only
if it manages to get things right. And this view that current science has
got it altogether right evidently has its problems. For science constantly
changes its mind, not just with regard to incidentals but even on very
fundamental issues. The history of science is the story of the replacement of one defective theory by another. So how can one plausibly
maintain a scientific realism geared to the idea that "science correctly
describes reality"?

The characteristic genius of scientific realism is inherent in its equating of the theory-creatures envisioned in current natural science with the
domain of what actually exists. But this equation would work only if
science, as it stands, has actually "got it right". And this is something we
are certainly not able – and not entitled – to claim.

All too clearly there is insufficient warrant for and little plausibility to the claim that the world is as our present-day science claims it to be – that *our* science is *correct* science and offers the definitive "last word" on the issues regarding its creatures-of-theory. We can learn by empirical inquiry about empirical inquiry itself. And one of the key things to be learnt is that at no actual stage does natural science yield a firm, final, unchanging result.

The ultimate untenability of scientific theories is one of the very few points of consensus of modern philosophy. When Karl Popper writes "From a rational point of view, we should not 'rely' on any (scientific) theory, for no theory has been shown to be true, or can be shown to be true . . ."[4], he speaks for the entire tradition of modern science scholarship from Charles Sanders Peirce to Nancy Cartwright. We must unhesitatingly presume that, as we manage to push our inquiries through to deeper levels, we will get a very different view of the constituents of nature and their laws. Its changeability is a fact *about* science that is as inductively well-established as any theory *of* science itself. Science is not a static system but a dynamic process.

If the future is anything like the past, if historical experience affords any sort of guidance in these matters, then we know that *all* of our presently favored scientific theses and theories will ultimately turn out to be untenable – that none are correct exactly as is. All the experience we can muster indicates that there is no justification for viewing our science as more than an inherently imperfect stage within an ongoing development. The ineliminable prospect of far-reaching future changes of mind in scientific matters destroys any prospect of claiming that the world is as we now claim it to be – that science's view of nature's constituents and laws is correct.

The postulation of the reality of science's commitments is viable only if done *provisionally*, in the spirit of "doing the best we can manage at present, in the current state-of-the-art". Our prized "scientific knowledge" is no more than our "current best estimate" of the matter. The step of reification is always to be taken provisionally, subject to a mental reservation of presumptive revisability.

We *think* we are bound to have encompassed the truth when we have "boxed the compass" of alternatives:

– Cancer is caused by a virus.
– Cancer is caused by something other than a virus.

Surely, we tell ourselves, the truth must lie either on the one side or the other. The *disjunction* of these possibilities must evidently be true.

But must it be true? Observe that both theses alike adopt a common presupposition. Both envision a "disease entity" designated by the term *cancer*. But what if there is no such *thing* as "cancer" because there just is not a stable entity, but merely a diversified family of ailments united only by Wittgensteinian "family resemblances"? What if we come to believe that the very concept at issue is not "objectively meaningful" – if cancer as such simply is not there to have a cause at all.

It is a presupposition of every factual statement that its concepts have a bearing on the real world, that they are indeed applicable to things, that nature actually exemplifies them, that they are "objectively meaningful". But as "magnetic effluxes" and the "luminiferous aether" show, this presupposition can be totally mistaken in scientific contexts.

The current state of "scientific knowledge" is simply one state among others that share the same imperfect footing of ultimate correctness or truth. The "science of the day" must be presumed inaccurate no matter what the calendar says. We unequivocally realize there is a strong prospect that we shall ultimately recognize many or most of our current scientific theories to be false and that what we proudly flaunt as "scientific knowledge" is a tissue of hypotheses – of tentatively adopted contentions many or most of which we will ultimately come to regard as quite untenable and in need of serious revision, or perhaps even abandonment. It is this fact that blocks the option of a scientific realism of any straightforward sort. Not only are we not in a position to claim that our knowledge of reality is *complete* (that we have gotten at the *whole* truth of things), but we are not even in a position to claim that our "knowledge" of reality is *correct* (that we have gotten at the *real* truth of things). Such a position calls for the humbling view that just as we think our predecessors of a hundred years ago had a fundamentally inadequate grasp on the "furniture of the world", so our successors of a hundred years hence will take the same view of our purported knowledge of things.

Thus, a clear distinction must be maintained between "*our conception of* reality" and "reality *as it really is*". We realize that there is little justification for holding that present-day natural science describes reality and depicts the world as it really is. And this constitutes a decisive impediment to straightforward realism. It must inevitably constrain and condition our attitude towards the natural mechanisms envisioned in contemporary science. We certainly do not – or should not – want to reify (hypostatize) the "theoretical entities" of current science, to say flatly and without qualification that the contrivances of *our* present-day science correctly depict the "furniture of the real world". We do not – or at any rate, given the realities of the case, should not – want to adopt categorically the ontological implications of scientific theorizing in just exactly the state-of-the-art configuration presently in hand. A realistic awareness of scientific fallibilism precludes the claim that the furnishings of the real world are exactly as our science states them to be – that electrons actually are just what the latest *Handbook of Physics* claims them to be.

We have to come to terms with the realism-impeding fact that our scientific knowledge of the world fails in crucial respects to give an accurate picture of it. Certainly, we subscribe for the most part to the working hypothesis that in the domain of factual inquiry *our* truth may be taken to be *the* truth. All the same, we realize that our science is not definitive, that reality is *not* actually as we currently picture it to be, that our truth is not the real truth, that we are probably quite wrong in supposing that the furnishings of "our science" actually exist exactly as it conceives them to be. No doubt "reality itself", whatever that may be, is real enough, but our "empirical reality" – reality as our science conceives it – is a fiction. Our scientific description of reality is a mind-devised, man-made artifact that cannot actually be accepted at face value.

Ultimately, when science is seen in historical perspective, it becomes clear that there is no adequate justification for thinking that natural science as we now have it is definitively correct.

However, what of a weaker realism to the effect that science is only correct in part? Even this suggestion has its difficulties for it immediately invites the question, *which* part? Just how are we to discriminate the correct from the incorrect in science as we have it, given our endorsement of the whole lot? This is a daunting challenge indeed!

3. FUTURE SCIENCE DOES NOT DO THE JOB

Perhaps, however, a different sort of realism can be maintained. Perhaps it is not *current* science that provides for the epistemic component of scientific realism but rather the science of the future. Despite our science's failure to characterize the real adequately, perhaps that of our successors will do so. Though this view is worth contemplating, its prospects are not auspicious. For the question arises: Just which future? After all, there is little reason to think that the status of tomorrow's science is in principle different from that of today's.

The equilibrium achieved by natural science at *any* given stage of its development is always an unstable one. Scientific theories have a finite lifespan. They come to be modified or replaced under various innovative pressures, in particular the enhancement of observational and experimental evidence (through improved techniques of experimentation, more powerful means of observation and detection, superior procedures for data-processing, etc.). And so, a "state-of-the-art" of natural science is a human artifact and like all other human creations has a finite lifespan. As something that comes into being within time, the passage of time will also bear it away. Clearly our present science simply is not in a position to deliver a definitive picture of physical reality. And there is no reason to think that, in the future, scientific theorizing must in principle reach a final and permanent result. Scientific work at the creative frontier of theoretical innovation is always done against the background of the realization that anybody's "findings" – one's own included! – will eventually be abandoned and become superseded by something rather different. Only the aims of natural science are stable, not its substantive questions – let alone its answers to them!

As C. S. Peirce and Karl Popper after him have insisted, we must acknowledge an inability to attain the final and definitive truth in the theoretical concerns of natural science – in particular at the level of theoretical physics. On all indications, we shall never be able to bring the scientific project to its decisive end and to obtain the final and definitive truth with regard to nature's ways. Our present world-view seems destined to crumble in the wake of ongoing scientific change – regardless of when the "present" at issue may fall. It seems to be the inevitable destiny of physics that its practitioners in every generation will see the theories of an earlier era as mistaken, as full of errors of omission and of commission as well. In all human probability, the physicists in the year 3000 will deem our physics no more correct than we deem that of a hundred years ago – and the same destiny of ultimate revision awaits their own views in turn.

If there is one thing we can learn from the history of science, it is that the scientific theorizing of one day is looked upon by the next as deficient. At *every* stage of its development, its practitioners, looking backwards with the wisdom of hindsight, will unquestionably view the work of their predecessors as seriously deficient and their theories as fundamentally inadequate in critical regards. There is no reason to see the posture of our successors as fundamentally different from our own in this respect. We know that science can be improved. But we also acknowledge that it cannot be perfected. Considerations of general principle, as well as the lessons of the history of science prevent our taking the stance that the world is as science depicts it to be – be it present-day science or the science of A.D. 3000. And this realization constitutes a decisive impediment to an eschatological realism that looks to future science for the finality that present science is unable to supply.

In the domain of natural science as we do – or ever shall – actually have it, access to anything identifiably final and definitive is denied us. Thus the definitively correct science is no more (though also no less) than an idealization. It represents an ideal which, like other ideals, is worthy of pursuit, despite the fact that we must recognize that its full attainment lies beyond our grasp. Scientific realism represents an over-optimistic idealization; it is not a position that is realistically tenable in any straightforward or unqualified way.

CHAPTER TWO

SCIENTIFIC PROGRESS AS NONCONVERGENT

SYNOPSIS. (1) Scientific progress is a matter of the increasingly far-reaching exploration of the "parameter-space" of our natural environment. (2) Scientific inquiry strives towards a systematization of theorizing conjecture with the data of experience. It involves the ongoing restoration of an equilibrium between data and theory through the destabilizing shocks of enlarged experience. The technologically mediated entry into new regions of parameter space confronts us with this task in ever-renewed forms. (3) In principle, at any rate, the prospect of ongoing "scientific revolutions" is potentially unending. (4) There is no warrant for adopting a theory of convergence that sees the innovations of theorizing science as being of constantly diminishing scope. Neither can a scientific realism rely on present science, nor are its difficulties resolved by looking to "the long run".

1. THE EXPLORATION MODEL AND ITS IMPLICATIONS

Even though neither present nor future science manage to depict reality adequately, perhaps there is a gradual *convergence* towards a true account of nature at the level of scientific theorizing. To evaluate this prospect critically, it is useful to invoke what might be called the "exploration model" of scientific progress.

In developing natural science, we began by exploring the world in our own locality – not just our *spatial* neighborhood but our *parametric* neighborhood in the "space" of physical variables like temperature, pressure, radiation, and so on. Near our natural "home-base" we are – thanks to the evolutionary heritage of our sensory and cognitive apparatus – able to operate with relative ease and freedom, scanning nature for data with the unassisted senses. But in due course we did all that we could manage in this way. To do more, we proceeded to extend our interactive probes into nature more deeply, deploying increasing technical sophistication to achieve increasingly high levels of capability. We have pushed ever further out from our evolutionary home-base in nature toward increasingly remote frontiers. From the egocentric standpoint of our local region of parameter space, we have journeyed ever further to "explore" in the manner of a "prospector" searching for cognitively significant phenomena along the various parametric dimensions.

This picture is clearly not one of *geographical* exploration, but of the *physical* exploration – and subsequent theoretical rationalization – of "phenomena" which are distributed over the "space" of the physical quantities spreading out all about us. This exploration-metaphor forms the basis of the conception of scientific *research* as a prospecting *search* for the new phenomena needed for significant new scientific findings. As the range of telescopes, the energy of particle accelerators, the effectiveness of low-temperature instrumentation, the potency of pressurization equipment, the power of vacuum-creating contrivances, and the accuracy of measurement apparatus increases – that is, as our capacity to move about in the parametric space of the physical world is enhanced – new phenomena come into our perception, with the result of enlarging the empirical basis of our knowledge of natural processes. The key to the great progress of contemporary physics lies in the enormous strides made in this regard.[1]

No doubt, nature is in itself uniform as regards the distribution of its diverse processes across the reaches of parameter space. It does not favor us by clustering them in our parametric vincinity: significant phenomena do not dry up outside our "neighborhood." But *cognitively* significant phenomena in fact become increasingly sparse because the scientific mind has the capacity to do so much so well early on. Our power of theoretical triangulation is so great that we can make a disproportionately effective use of the phenomena located in our local parametric neighborhood. But scientific innovation becomes more and more difficult – and expensive – as we push out farther and farther from our evolutionary home-base toward ever more remote frontiers. After the major findings accessible at a given data-technology level have been achieved, further major findings become realizable when one ascends to the next level of sophistication in data-relevant technology. We confront a situation of technological escalation. The need for new data constrains looking further and further from man's familiar "home-base" in the parametric space of nature. Thus, while further significant scientific progress is in principle always possible – there being no absolute or intrinsic limits to significantly novel facts – the *realization* of this ongoing prospect for scientific discovery demands a continual enhancement in the technological state-of-the-art of data extraction or exploitation. It is of the very essence of the enterprise that natural science is forced to press into ever remoter reaches of parametric space.

This ever more far-ranging exploration of the spectra associated with different conditions in nature demands continual increases in physical power. To enable our experimental apparatus to achieve greater velocities, higher frequencies, lower or higher temperatures, greater pressures, larger energy-excitations, stabler conditions, greater resolving power, etc., requires ever more powerful equipment capable of continually more enhanced performance. The *sort* of "power" at issue will, of course, vary with the nature of the particular parametric dimension under consideration – be it velocity, frequency, temperature, etc. – but the general principle remains the same. Scientific progress depends crucially and unavoidably on our technical capability to penetrate into the increasingly distant – and increasingly difficult – reaches of the power-complexity spectrum of physical parameters, to explore and to explain the ever more remote phenomena encountered there. Only by operating under new and previously inaccessible conditions of observational or experimental interactions with nature – attaining ever more extreme temperature, pressure, particle velocity, field strength, and so on – can we bring new impetus to scientific progress. However, it is useful and important to emphasize that this "parametric space" parametrizes the degree of *capability on our part* to realize conditions at variance from our evolutionary "home-base", and not simply positions that lie one-sidedly in the constitution of nature *per se*.

This idea of the "exploration" of parametric space provides a basic tool for the present model of the mechanism of scientific innovation in mature science. New technology increases the range of access within the parametric space of physical processes. Such increased access brings new phenomena to light, and the detection, scrutiny, and theoretical systematization of these phenomena is the basis for growth in our scientific understanding of nature. As an army marches on its "stomach", (its logistical support), so science depends upon its "eyes" – it is crucially dependent on the technological instrumentalities which constitute the sources of its data. Natural science is fundamentally *empirical*, and its advance is critically dependent not on human ingenuity alone, but on the monitoring observations to which we can only gain access through interactions with nature. The days are long past when useful scientific data can be gathered by unaided sensory observation of the ordinary course of nature. *Artifice* has become an indispensable route to the acquisition and processing of scientifically useful data: the sorts of data on which scientific discovery nowadays depends can only be generated by technological means.

Without an ever-developing technology of experimentation and observation, scientific progress would grind to a halt. The discoveries of today cannot be attained with yesterday's instrumentation and techniques. To secure new observations, to test new hypotheses, and to detect new phenomena, an ever more powerful technology of inquiry is needed. Throughout the natural sciences, technological progress is a crucial requisite for cognitive progress. We are embarked on an endless endeavor to improve the range of effective observational and experimental intervention. Only by operating under new and previously inaccessible conditions – attaining extreme temperature, pressure, particle velocity, field strength, and so on – can we realize those circumstances that enable us to put our hypotheses and theories to the test. As an acute observer has rightly remarked: "Most critical experiments [in physics] planned today, if they had to be constrained within the technology of even ten years ago, would be seriously compromised."[2]

This situation points toward the idea of "technological levels", corresponding to the successive state-of-the-art stages in the technology of inquiry with regard to data-generation and processing – giving rise to successively "later generations" of investigative instrumentation and machinery. These levels are generally separated from one another by substantial (order-of-magnitude) improvements in performance with regard to such information-providing parameters as measurement exactness, data-processing volume, detection-sensitivity, high voltages, high or low temperatures, and so on.

Frontier research is true *pioneering*: what counts is not just doing it but doing it *for the first time*. Aside from the initial reproduction of claimed results needed to establish their reproducibility of results, repetition in *research* is generally pointless. As one acute observer has remarked, scientific technology becomes diffused "from the research desk down to the schoolroom".

The emanation electroscope was a device invented at the turn of the century to measure the rate at which a gas such as thorium loses its radioactivity. For a number of years it seems to have been used only in the research laboratory. It came into use in instructing graduate students in the mid-1930s, and in college courses by 1949. For the last few years a cheap commercial model has existed and is beginning to be introduced into high school courses. In a sense, this is a victory for good practice; but it also summarizes the sad state of scientific education to note that in the research laboratory itself the emanation electroscope has long since been removed from the desk to the attic.[3]

The enormous power, sensitivity, and complexity deployed in present-day experimental science have not been sought for their own sake but rather because the research frontier has moved on into an area where this sophistication is the indispensable requisite of ongoing progress. In science, as in war, the battles of the present cannot be fought effectively with the armaments of the past.

The salient characteristic of this situation is that once the major findings accessible at a given data-technology level have been attained, further major progress in the problem-area requires ascent to a higher level on the technological scale. Every data-technology level is subject to *discovery-saturation*, but the exhaustion of prospects at a given level does not, of course, bring progress to a stop. For after the major findings accessible at a given data-technology level have been realized, further major findings become realizable by ascending to the next level of sophistication in data-relevant technology.

We arrive therefore at a situation of *technological escalation*. The need for new data forces us to look further and further from man's familiar "home-base" in the parametric space of nature. Thus, while scientific progress is in principle always possible – there being no absolute or intrinsic limits to significant scientific discovery – the *realization* of this ongoing prospect demands a continual enhancement in the technological state-of-the-art of data extraction or exploitation.

2. THEORIZING AS INDUCTIVE PROJECTION

Scientific theorizing is a matter of triangulation from observations – of inductive generalization from the data. And (sensibly enough) induction constructs the most economical structures to house these data comfortly. It discerns the simplest overall pattern of regularity that can adequately accommodate this data regarding cases-in-hand, and then projects them across the entire spectrum of possibilities in order to answer our general questions. As a fundamentally inductive process, scientific theorizing involves the search for, or the construction of, the least complex theory-structure capable of accommodating the available body of data. Induction is a matter of projecting beyond the data as is necessary to get answers while staying as close to the data as possible, all the while proceeding under the aegis of established principles of inductive systematization, uniformity, simplicity, harmony, and the rest that implement the general idea of cognitive economy. The assurance of inductively authorized contentions turns exactly on this issue of tightness of fit: of consilience, mutual interconnection, and systemic enmeshment. Induction is a matter of building up the simplest structure capable of "doing the job". The key principle is that of simplicity and the ruling injunction that of cognitive economy. Complications cannot be ruled out, but they must always pay their way in terms of increased systemic adequacy!

Simplicity and generality are the cornerstones of cognitive systematization. But one very important point must be stressed in this connection. The idea of a *coordinative systematization of question-resolving conjecture with the data of experience* may sound like a very conservative process. This impression would be quite incorrect. The drive to systematization embodies an imperative to broaden the range of our experience – to extend and to expand the data-base from which our theoretical triangulations proceed – which is no less crucial than assuming their elegance. Simplicity/harmony and comprehensiveness/inclusiveness are two components of one whole. That is why the ever-widening exploration of nature's parameter space is an indispensable part of the process.

With the enhancement of technology the size of this body of data inevitably grows. Technological progress constantly enlarges the "window" through which we look out upon parametric space. In developing natural science we use this window of capability to scrutinize parametric space, continually augmenting our data-base and then generalizing upon what we see. And what we have here is not a lunar landscape where once we have seen one sector, we have seen it all, and where theory-projections from lesser data generally remain in place when further data comes our way. Historical experience shows that there is every reason to expect that our ideas about nature are subject to constant radical changes as we "explore" parametric space more extensively. The technologically mediated entry into new regions of parameter space constantly destablilizes the attained equilibrium between data and theory.

At this point the situation contemplated in the crude curve-fitting illustration of Figure 1 becomes instructive. Our exploration of physical parameter space is inevitably incomplete. We can never exhaust the whole range of temperatures, pressures, particle velocities, etc. And so, we inevitably face the (very real) prospect that the regularity structure of the, as yet, inaccessible cases generally does not conform to the

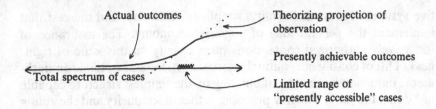

Fig. 1. Problems of extrapolation.

patterns of regularity prevailing in the presently accessible cases. By and large, future data do not accommodate themselves to present theories. Newtonian calculations worked marvelously for predicting solar-system phenomenology (eclipses, planetary conjunctions, and the rest) but this does not show that classical physics is free of any need for fundamental revision.

Even extraordinary accuracy with respect to the entire range of *currently manageable* cases does not betoken actual correctness – it merely reflects adequacy over that limited range. And no matter how far we broaden that "limited range of 'presently accessible' cases", we still achieve no assurance (or even probability) that a theory-corpus which accommodates (perfectly well) the range of "presently achievable outcomes" will hold "across-the-board".

Scientific theory-formation is, in general, a matter of spotting a local regularity of phenomena in parametric space and then projecting it "across-the-board", maintaining it globally. The theoretical claims of science are themselves never local – they are not spatiotemporally local and they are not parametrically local either. They stipulate how things are always and everywhere. And so it does not require a sophisticated knowledge of statistics to realize that inductive projection of the sort we make in science is invariably a risky enterprise. And it does not require a sophisticated knowledge of history of science to realise that our worst fears are usually realized – that it is seldom if ever the case that our theories survive intact in the wake of extensions in our access to sectors of parametric space. The history of science is a history of episodes of leaping to the wrong conclusions.

In pursuing the venture of scientific inquiry, we scan nature for interesting phenomena and for the explanatorily useful regularities suggested by them. At each stage we try to embed the phenomena and their regularities within the simplest (cognitively most efficient) explanatory fabric to answer our questions about the world and to guide our interactions in it. Breadth of coverage and economy of means are our guiding stars.

Against this background let us contemplate an analogy. Let us suppose that we investigate some domain of phenomena on such a basis, and that in the first instance the picture we arrive at is one showing a certain sort of regularity as shown below.

We say: "Aha, this sector of the world's processes proceeds in the manner of zigzags." But at the next level we investigate those zigzags more closely. We note now that they have the following form.

We say: "We did not quite have it right to begin with. This sector of the world's processes actually has the character of castellations." And so, at the next level we investigate those castellations more closely. We now note that they in turn have changed form.

We now say: "Aha, this sector of the world is made up of configurated waves." And so this form of observation continues at every successive level of technological "magnification" in our interaction with physical nature, as her apparent *modus operandi* looks very different. At each level, the "governing regularities" of nature take on an aspect very different from what went before and crucially disparate from it.

Note, however, that at each stage we can readily comprehend and explain what went before. We can always say, "Yes, of course, given that that is how things stand, it is quite understandable that earlier on, when we went at it in a certain way, we arrived at the sort of results we did – wrong though they are." But, of course, this wisdom is one of hindsight only. At no stage do we have the prospect of using *foresight* to predict what lies ahead. The impossibility of foreseeing the new phenomena that awaits us means that at no point can we prejudge what lies further down the explanatory road.

Let us now turn from a concern with the *lawful comportment* of the world's phenomena to the *constitution* of its things. An analogy will once again be helpful. Suppose when we initially investigate a type of object X, proceeding to the first level of sophistication we see it is constituted of parts whose structure is O-like. On closer investigation (at the next level of sophistication), however, we see that these "component parts" were not actually units, but mere constellations, mere clouds of small specks as per ░. When we investigate further, it emerges that the component specks that constitute these "clouds" themselves have the rectangular form ▢. And suppose further that at the next level those rectangular configurated "components" themselves emerge as mere constellations, composed of triangular constituents of the form △, and so on. As this analogy indicates, physical nature exhibits a very different aspect when viewed from the vantage point of different levels of sophistication in the technology of nature-investigator interaction.

Thus both as regards the observable *regularities* of nature and the discernible *constituents* of nature, very different results emerge at the various levels of the observational state-of-the-art. At every stage we deal with a different order or aspect of things. And the reason why nature exhibits different aspects at different levels is not that nature herself is somehow stratified and has different levels of being or of operation but rather that (1) the character of the available nature-investigative interaction is variable and differs from level to level, and (2) the character of the "findings" at which one arrives will hinge on the character of these nature-investigative interactions. What we "find" in some degree "lies in the 'eyes' of the beholder" in reflecting the apposite technology of observation. What we detect or "find" in nature is always something that depends on the mechanisms by which we search. The phenomena we detect will depend not merely on nature's doings alone, but on the physical and conceptual instruments we use in probing nature.

As Bacon saw, nature will never tell us more than we can forcibly extract from her. And what we can manage to extract at the successive levels of interaction made accessible by successively deeper probes will wear a steadily changing aspect.

Given that we can only learn about nature by interacting with it,

Newton's third law of countervailing action and reaction becomes a fundamental principle of epistemology. Everything depends on just how *and how hard* we can push against nature in situations of observational and detectional interaction. And we cannot "get to the bottom of it" where nature is concerned. Nature always has hidden reserves of power.

Successive stages in the technological state-of-the-art of scientific inquiry lead us to different views about the nature of things and the character of their laws. But the sequence of successively more powerful and sophisticated instrumentalities on the side of inquirers need not be matched by any coordinated succession of layers in the constitution of physical existence somehow captured "correctly" by our inquiry at corresponding levels of sophistication. The "layers" we encounter principally reflect our own procedures.

Accordingly, it is a wholly unwarranted supposition that there is a sequence of nature-levels placed conveniently alongside our inquiry levels, in a parallel coordination, that makes for an elegantly ladder-like configuration. Nature just goes along "doing her thing". Nature has no layers, no differentiated physical strata or levels.[4] The only "layers" now in view are those at issue with the technology of "probing" – of observation and manipulation – that is used by certain sorts of beings in the course of their *interaction* with nature. (Nature no more has levels because she reacts differently to different sorts of probes than does an animal which responds differently to different stimuli.) The only physical layers are process-relative, hinging on the character of our modes of observation and manipulation.

3. SCIENTIFIC REVOLUTIONS AS POTENTIALLY UNENDING

This perspective on the development of science has important implications. It means that science cannot be a complete *system*, a finished structure of knowledge, but is and will ever remain a *process* – an inquiring activity whose ultimate goal may be the completion of a finished and perfected system, but which proceeds in the full recognition that this aim is ultimately unreachable. We have to accept the idea that while *progressing* makes sense, *arriving* does not. We cannot attain perfection; we can always do more, but we cannot "do it all". Neither theoretical issues of general principle nor the actualities of historical experience suggest that scientific progress need ever come to a stop.[5] Of course, it is possible that for reasons of exhaustion, of penury , or of discouragement, we humans might cease to push the frontiers forward. But should we ever abandon the journey, it will be for reasons such as these and not because we have reached the end of the road.

Some theorists regard science as an essentially closed venture that will ultimately come to the end of its tether. They see the scientific project as a potentially bounded venture, subject to the idea that in scientific inquiry as in geographical exploration, we are ultimately bound to arrive at the end of the road.[6] But this position is eminently problematic. For there is good reason to think of nature as cognitively inexhaustible: we can, in theory, always learn more and more about it, attaining ever new horizons of discovery, with the new no less interesting or significant than the old.

But how can unlimited scientific discovery be possible? To underwrite the prospect of endless progress in the discovery of natural laws, some theorists have felt compelled to stipulate an intrinsic infinitude in the structural makeup of nature itself.[7] The physicist David Bohm, for example, tells us "at least as a working hypothesis science assumes the infinity of nature; and this assumption fits the facts much better than any other point of view that we know."[8] Bohm and his congeners thus postulate an infinite quantitative scope or an infinite qualitative diversity in nature, assuming either a principle of unending intricacy in its makeup or one of unending orders of spatiostructural nesting. But is this sort of thing needed at all? Does the prospect of potentially limitless scientific progress actually require structural infinitude in the physical composition of nature along some such lines? The answer is surely negative.

The prime task of science lies in discovering the laws of nature, and it is law-complexity that is crucial for this purpose.[9] Even a mechanism of finitely ramified structure can have endlessly complex laws of operation. For if one remains at a fixed level of scale in physical magnitude, one can still have rearrangements of items and rearrangements of rearrangements *ad infinitum*, with emergent lawful characteristics arising at every stage. The workings of a structurally finite and indeed simple system can yet exhibit an infinite intricacy in *operational or functional complexity*, exhibiting this limitless complexity in its workings rather than at the spatio-structural or compositional level. While the number of constituents of nature may be small, the ways in which they may be combined can be infinite. Think of the examples of letters, syllables, words, sentences, paragraphs, different sorts of books, libraries, library systems. There is no need to assume a "ceiling" to such a sequence of levels of integrative complexity. The emergence of new concept-concatenations and new laws can be expected at every stage. Each level exhibits its own order. The laws we attain at the n-th level can have features whose investigation lift us to the (n + 1)st. New phenomena and new laws can arise at every new level of integrative order. Knowing the frequency with which individual letters like A and T occur in a text will not tell us much about the frequency with which a combination such as AT occurs. When we change the purview of our conceptual horizons, there is always, in principle, more to be learned. The different facets of nature can generate new strata of laws that yield a potentially unending sequence of levels, each giving rise to its own characteristic principles of organization, themselves quite unpredictable from the standpoint of the other levels.

A "working hypothesis" of the *structural infinitude* of nature is not needed to assure non-terminating progress in science. An unending depth in the operational or functional complexity of nature would be quite enough to underwrite the potential limitlessness of science.

But even this is too much. Actually, neither unending *structural* nor *operational* complexity is required to provide for cognitive inexhaustibility. The usual recourse to an infinity-of-nature principle is strictly one-sided, placing the burden of responsibility for the endlessness of science solely on the shoulders of nature herself. According to this view, the potential endlessness of scientific progress requires limitlessness on the side of its *objects*, so that the infinitude of nature must be postulated either at the structural or at the functional levels. But this is a mistake.

Science, the cognitive exploration of the ways of the world, is a matter of the *interaction* of the mind with nature – of the *mind's exploitation of the data to which it gains access* in order to penetrate the "secrets of nature". The crucial fact is that scientific progress hinges not just on the structure of nature but also on the structure of the information-acquiring processes by which we investigate it.

Ongoing cognitive innovation thus need not be provided for by assuming (as a "working hypothesis" or otherwise) that the system being investigated is infinitely complex in its physical or functional make-up. It suffices to hypothesize an endlessly ongoing prospect of securing fuller information about it. The salient point is that it is *cognitive* rather than structural or operational complexity that is the key here. After all, even when a scene is itself only finitely complex, an ever ampler view of it will come to realization as the resolving power of our conceptual and observational instruments is increased. And so, responsibility for the open-endedness of science need not lie on the side of nature at all but can rest one-sidedly with us, its explorers.

When we make measurements to accuracy A, the world may appear X_1-wise; and when to accuracy $(1/2)A$ it may appear X_2-wise; and when to accuracy $(1/2^n)A$ it may appear X_n-wise. At each successive state-of-the-art stage of increased precision in our investigative proceedings, the world may take on a very different nomic appearance, not because it *changes*, but simply because at each state it *presents* itself differently to us.

Accordingly, the question of the ongoing progressiveness of science should not be confined to a consideration of nature alone, since the character of our information-gathering procedures, as channeled through our theoretical perspectives, is also bound to play a crucial part.[10] Innovations on the side of data can generate new theories, and new theories can transform the very meaning of the old data. This dialectical process of successive feedback has no inherent limits, and suffices to underwrite a prospect of ongoing innovation. Even a finite nature can, like a typewriter with a limited keyboard, yield an endlessly varied text. It can produce a steady stream of *new* data – "new" not necessarily in kind but in their functional interrelationships and thus in their theoretical implications – on the basis of which our knowledge of nature's operative laws is continually enhanced and deepened.

These various considerations combine to indicate that *an assumption of the quantitative infinity of the physical extent of the natural universe or of the qualitative infinity of its structural complexity is simply not required to provide for the prospects of ongoing scientific progress*. Ongoing discovery is as much a matter of how we inquirers proceed with our work as it is of the object of inquiry itself. And this fact constrains us to recognize that even a finitely complex nature can provide the domain for a virtually endless course of new and significant discovery. There is no good reason to think that the natural science of a finite world is an inherently closed and terminable venture, and no adequate basis for the view that the search for greater "depth" in our understanding must eventually terminate at a logical end.[11] On all indications, historical as well as theoretical, the prospect of ongoing "scientific revolutions" is potentially unending.

4. IS LATER LESSER?

To salvage scientific realism in the face of the prospect of unending scientific change, it may be tempting to abandon the idea of an ultimate state of completed science and adopt Charles Sanders Peirce's idea of *convergent approximation*.[12] This calls for envisaging a situation where, with the passage of time, the results we reach grow increasingly concordant and the outcomes attained become less and less differentiated. In the face of such a course of successive changes of ever-diminishing significance, we could proceed to maintain that the world really is not as *present* science claims it to be, but rather is as the ever more clearly emerging science-in-the-limit claims it to be. The reality of ongoing change is now irrelevant as with the passage of time the changes matter less and less. We increasingly approximate an essentially stable picture. This prospect is certainly a theoretically possible one. But neither historical experience nor considerations of general principle provide reason to think that it is actually possible. Instead, quite the reverse!

Any theory of convergence in science, however carefully crafted, will shatter against the *conceptual innovation* that continually brings entirely new, radically different scientific concepts to the fore, carrying in its wake an ongoing wholesale revision of "established fact". Investigators of an earlier era not only did not *know* what the half-life of californium was, but they would not have *understood* it even if this fact had been explained to them. This aspect of the matter deserves closer attention.

"Which of the four elements (air, earth, fire, water) is the paramount *'arche'*, the fundamental type of stuff from which the the whole of physical reality originates?" asked the early Milesians. They contemplated just those four alternatives together with the fifth possibility of a neutral, intermediate stuff. It did not occur to them that their whole inquiry was abortive because it was based on a misguided conception of "elements". Nor did it appear to be a realistic prospect to all those late nineteenth-century physicists who investigated the properties of the luminiferous aether that no such medium for the transmission of light and electromagnetism might exist at all.

In factual inquiry into the ways of the world we can do no better than to pose questions and canvass the currently visible alternatives. But the questions we can pose are limited by our conceptual horizons. And the answers we can envision are also limited by the cognitive "state-of-the-art". (The Greeks could not have asked about continental drift; the Romans could not have thought of explaining the tides through gravitation.) And, of course, the whole process of canvassing answers can come to grief because the very question being asked is based on untenable suppositions.

Ongoing scientific progress is not simply a matter of increasing accuracy by extending the numbers in our otherwise stable descriptions of nature out to a few more decimal places. Significant scientific progress is genuinely revolutionary in involving a *fundamental change of mind* about how things happen in the world. Progress of this caliber is generally a matter not of adding further facts – on the order of filling in a crossword puzzle – but of changing the framework itself. And this fact blocks the theory of convergence.

In any convergent process, later is lesser. But since scientific progress on matters of fundamental importance is generally a matter of replacement rather than mere supplementation, there is no reason to see the *later* issues of science as *lesser* issues in the significance of their bearing upon science as a cognitive enterprise – to think that nature will be cooperative in always yielding its most important secrets early on and reserving nothing but the relatively insignificant for later on. (Nor does it seem plausible to think of nature as perverse, leading us ever more deeply into deception as inquiry proceeds.) A very small scale effect – even one that lies very far out along the extremes of a "range exploration" in terms of temperature, pressure, velocity, or the like – can force a far-reaching revolution and have a profound impact by way of major theoretical revisions. (Think of special relativity in relation to aether-drift experimentation, or general relativity in relation to the perihelion of Mercury.)

Given that natural science progresses mainly by substitutions and replacements that go back to first principles and lead to comprehensive overall revisions of our picture of the phenomena at issue, it seems sensible to say that the shifts across successive scientific "revolutions" maintain the same level of overall significance. At the cognitive level, a scientific innovation is simply a matter of change; it is neither a convergent nor a divergent process.[13]

On this basis, one arrives at a view of scientific progress as a steady-state condition in which every major successive stage in the evolution of science yields innovations, and innovations of roughly equal *overall* interest and importance. Accordingly, there is little alternative but to reject convergentism as a position that lacks the support not only of considerations of general principle but also of the actual realities of our experience in the history of science.[14]

These deliberations have important bearings on scientific realism. They indicate that scientific realism cannot take an eschatological turn. We cannot say that "the real truth" lies with *current* science, nor yet with *future* sciences, nor yet with the *emergent* science of a developing convergence. What, then, *can* we say in this regard?

CHAPTER THREE

IDEAL-SCIENCE REALISM

SYNOPSIS. (1) Only ideal or perfected science accurately and reliably depicts reality, and not science as we do or shall actually have it. (2) In matters of inductive theorizing, "the actual truth" is attained only in the *ideal* limit. (3) In the course of actual practice, scientific inquiry provides no more than our best available *estimate* of the truth. The only unproblematically viable sort of scientific realism is accordingly one that is geared to the ideal state.

1. REALITY IS ADEQUATELY DESCRIBED ONLY BY IDEAL SCIENCE, WHICH IS SOMETHING WE DO NOT HAVE

Once we "distance" ourselves from the cognitive commitments of our science by recognizing that they can and frequently do go awry, we must also acknowledge that "our scientific picture" of reality is not fully accurate, admitting that we have neither the inclination nor the warrant for claiming that reality actually is as it is purported to be by the science of the day. As concerns our cognitive endeavors, "man proposes and nature disposes," and it does so in both senses of the term: it disposes *over* our current scientific view of reality and it will doubtless eventually dispose *of* it as well. Given this circumstance, we have little alternative but to presume reality to have a character regarding which we are only imperfectly informed by natural science.

Success in providing a definitive truth about nature's ways is doubtless a matter of intent rather than one of accomplishment. Correctness in the characterization of nature is achieved not by *our* science, but only by *perfected* or *ideal* science – only by that (ineradicably hypothetical) state of science in which the cognitive goals of the scientific enterprise are fully and reliably realized. We are constrained to acknowledge that it is not *present* science, nor even *future* or *ultimate* science, but only *ideal* science that correctly describes reality – an ideal science that we shall never in fact attain, since it exists only in utopia and not in this mundane dispensation. Scientific realism must thus come to terms with the realization that reality is depicted by *ideal* (or perfected or "completed") science, and not by the real science of the day, which, after all, is the only one we have actually got – now or ever. Our science is constituted

26

of *putative* knowledge that does no more than to envision the truth as best we can discern it with the limited means at our disposal. Someone might object:

How can you maintain a clear line of distinction between real and merely putative truth or knowledge in science? After all, might it not possibly (perhaps even probably) happen that something that is merely conjectured, suspected, or estimated to be true will actually turn out to be so?

The proper response is clearly: "Of course it can". But this is beside the point. The distinction at issue is one that pertains not to the *content* of our claims, but to their *epistemic status*. Though it is possible that we have hit upon the actual (definitive) truth, we clearly cannot rely upon it. Even were it so, we could not *establish* it.

We must maintain a certain tentative and provisional stance towards our own scientific "knowledge". We fully realize that what we *take* to be true or real here is not always true or real. It is just this consideration that constrains us to operate with the distinction between "our putative reality" and "reality as such". We realize that what we think to be so – be it in science or in common life – frequently just is not so. We certainly cannot identify our achieved putative scientific truth with the real truth of the matter. No route save idealization is able to effect a sure and general connection between belief and the real truth. Only ideal or perfected science accurately and correctly depicts reality, and not science as we actually have it here and now. From the standpoint of epistemic status, truth is clearly an idealization – not what we *do* (or ever *will*) *have*, but what we *could have if* all the returns were in. It is thus in order to take a closer look at this matter of cognitive idealization.

2. SCIENTIFIC TRUTH AS AN IDEALIZATION

The history of science shows that our "discoveries" secured by way of the inductive coherentism of the scientific method constantly require adjustment, correction, replacement. We cannot say that our inductive inquiries about how things work in the world provide us with the real (definitive) truth, but rather that they provide us with *the best estimate* of the truth we can achieve in the circumstances to hand. Only at the idealized level of perfected science could we count on securing the real truth about the world that "corresponds to reality" as the traditional phrase has it.

The concept of science perfected – of an ideal and completed science that captures "the real truth" of things and satisfies all of our cognitive ideals (definitiveness, completeness, unity, consistency, etc.) – is at best a useful fiction, a creature of the imagination and not the secured product of inquirying rèason. This "ideal science" is, as its very name suggests, an idealization. We can only do the best we can in the cognitive state-of-the-art to *estimate* "the correct" answer to our scientific questions, which must suffice us because it is *all* that we can do. We recognize, or at any rate have little alternative but to suppose, that reality exists, accepting that there is such a thing as "the real truth" about the mind-independently real things of this world. But we are not in a position to state any final and definitive claims as to just exactly what it is like. Here we are confined to the level of plausible conjecture – of estimation.

Committed to the unproblematic claim *that* reality exists, we are, nevertheless, equally committed to the supposition that its nature is, in various not unimportant ways, different from what we think it to be. We can make no assured claims for our present-day science in this matter of "describing reality": the most we can do is to see it as affording our very best estimate of nature's descriptive constitution. We realize that science as it stands does not give us "definitive knowledge". We know that we will eventually come to see with the wisdom of hindsight that each of the claims of current frontier science, taken literally in the fullness of current understandings and explanations, is strictly speaking false.[1] The realities of the situation force us to accept the presumptive falsity of the claims made at the scientific frontier of the present day.

Accordingly, we have no alternative but to presume that *our* science as it currently stands does not present the real truth. All we can and should say is that current science affords us the *best estimate* of nature's ways that we can make here and now. "Our truth" in matters of scientific theorizing is not – and may well never actually be – the real truth. However confidently science may affirm its conclusions, its declarations are effectively provisional and tentative, subject to revision and even to outright abandonment and replacement. We must presume that science cannot attain an omega-condition of final perfection. The prospect of fundamental changes lying just around the corner can never be eliminated finally and decisively.

Inductive inquiry is truth-estimation. And here, as elsewhere, the gap between the real and the ideal must be acknowledged. What inquiry provides is "our purported truth" as contradistinguishable from "the real truth itself". The idea of "the definitive truth" functions as a regulative conception for us. It characterizes what we ideally aim at rather than what we actually obtain; it guides the direction of inquiry rather than describing its achievements.

In scientific matters we are never in a position to claim definitive truth with dogmatic certainty. The most we can ever realistically do is to claim what we do have as being the very best that one can possibly obtain in the circumstances. Scientific progress is *not* of a character that encourages us to reify (hypostatize) the theory-objects of science *as presently conceived* – regardless of the date the calendar may show. Once we have taken a realistic look at the history of science, it is scarcely an appealing proposition to maintain that *our* science, as it stands here and now, depicts reality actually and correctly – at best one can say that it affords an *estimate* of it that will doubtless stand in need of eventual revision. Its creatures-of-theory may in the final analysis not be real at all in the form in which the theory envisions them. This feature of science must crucially constrain our attitude towards its deliverances.

One recent commentator maintains the view that science's aim regarding true theories "leads to the view that science represents a utopian, and therefore irrational activity whose *telos* is, to the best of our knowledge, forever beyond our grasp."[2] But this position is profoundly wrong. It fails to deal appropriately with the standard gap between aspiration and attainment. In the practical sphere – in craftsmanship, for example, or our health care – we may *strive* for perfection, but cannot ever claim to have *attained* it. And the situation of inquiry is exactly parallel with what we encounter in other domains – ethics specifically included. The value of a goal, even of one that is not realizable, lies not in the benefits of its attainment (obviously and *ex hypothesi!*), but in the benefits that accrue from its pursuit. The view that it is rational to pursue a goal only if we are in a position to achieve its attainment or approximation is a mistaken one. The goal can be perfectly valid, and entirely rational if the indirect benefits of its adoption and pursuit are sufficient – if in striving after it we realize relevant advantages to a substantial degree. An unattainable ideal can be enormously productive.

This is not, of course, any reason to abandon the link to truth at the purposive level of the aims, goals and aspirations of science. The pursuit of scientific truth, or for that matter any other ideal in life, is not vitiated by the consideration that its full realization is not achievable.

The idea of definitive finality in scientific inquiry is more than an idealization. The conception of capital-T for Truth thus serves a negative and fundamentally regulative role marking the fact that whatever we have actually attained falls short of realizing our cognitive aspirations. Definitive truth is not something available which we actually claim to have in hand, but marks a fundamental contrast that *regulates* how we do and must view our claims to have got at the truth of things. It plays a role somewhat reminiscent of the functionary who reminded the Roman emperor of his mortality, in admonishing us that our pretentions to truth are always vulnerable.

Ideal science is not something we have got in hand here and now. And it is emphatically not something towards which we are moving along the asymptotic and approximative lines envisioned by Charles Sanders Peirce.[3] For Peirce identifies ideal or perfected science with an ultimate condition of science that is "fated" to emerge in the eventual course of history. But there is, of course, no guarantee of this whatsoever. Perfected science is not "what will emerge when" but "what would emerge if" – where a lot of (realistically unachievable) conditions must be supplied. As far as the actual course of history goes, we must recognize that even if it made sense to contemplate the Peircean idea of an eventual completion of science, there would be no guarantee that this completed science (given it existed!) would satisfy the definitive requirements of *perfected* science. Peircean convergentism is geared to the supposition that ultimate science – the science of the very distant future – will somehow prove to be an ideal or perfected science freed from the sorts of imperfections that afflict its predecessors. But the potential gap that arises here can only be closed by metaphysical assumptions of a most problematic sort.[4]

Existing science does not and presumably never will embody to perfection cognitive ideals of definitiveness, completeness, unity, consistency, etc. These factors represent an aspiration rather than a coming reality: a *telos* or direction rather than a realizable condition of things. Accordingly, there is no warrant for identifying *ideal* or perfected science with *ultimate* science. Perfected science is not something that exists here and how, nor is it something that lies ahead at some eventual offing in the remote future. It is not a real thing to be met with in this world. It is an

idealization that exists "outside time" – i.e., cannot attain actual existence at all. It lies outside history as a useful contrast-case that cannot be
numbered among the achieved realities of this imperfect world.

3. IDEAL-STATE REALISM AS THE ONLY VIABLE OPTION

There is only one world in existence: the real world as it actually is. But
we will not be able to say just what it is really like until the day when
natural science has been completed and perfected, which is to say *never*.
We must pursue the cognitive enterprise amid the harsh realities and
complexities of an imperfect world. And this means that what we
achieve in scientific inquiry is not the definitive truth as such, but only
our best estimate of it. In forming a just appreciation of our scientific
claims, the irremovable gap between the real and the ideal must once
again be acknowledged.

The thesis that "science truly describes the real world" must be
looked upon as a matter of intent rather than as an accomplished fact, of
aspiration rather than achievement, of the ideal rather than the real
state of things. Scientific realism is tenable only when it is the *ideal* state
of science that is at issue. (That, *ex hypothesi*, is what makes that state
into an ideal one.) But ideal-state realism, while demonstrably correct,
avails us less than we would like – we who occupy the sub-optimally real
rather than the perfected ideal order of things.

It is ideal science alone that gets at the definitive truth of things to
which authentic reality corresponds. Scientific realism is a viable position only with respect to that idealized science which, as we realize, we
do not now have – regardless of the "now" at issue. The only sort of
scientific realism that is unproblematically viable is an ideal-science
realism. We cannot be unqualified scientific realists or rather, ironically,
we can be so only in an idealistic manner, namely with respect to an
"ideal science" that we can never actually claim to possess.

Perfected science is an idealization – as is the scientific realism that
comes automatically in its wake. Now an ideal is not something we
encounter in experience, but rather the hypothetical projection or
extrapolation of what we encounter in experience. And the legitimacy
of our cognitive ideals as regulators inheres in their *utility* as guides to
inquiry, and specifically in their capacity to guide our thoughts and
efforts in constructive and productive directions. It is a fallacy to see the

validity of goals and ideals as residing solely in the presumed consequence of their *realization*. Their validation may reside not in arriving but
in the benefits we realize in the course of the pursuit itself. The striving
after an ideal science that affords us "the ultimate truth" about the
workings of nature seems to be a *telos* of just this sort. (We arrive at the
perhaps peculiar posture of an invocation of practical utility for the
validation of an ideal.[5])

CHAPTER FOUR

AGAINST INSTRUMENTALISM: REALISM AND THE TASK OF SCIENCE

SYNOPSIS. (1) Our inability to claim that natural science as we do or can have it depicts reality correctly must not be taken to mean that science is just a practical device – a mere instrument for prediction and control that has no bearing on describing "the nature of things". This sort of view would ride roughshed over the descriptive aims of the scientific enterprise. (2) For while we must accept a fallibilistic view of science, the fact remains that the quintessentially cognitive aspiration of getting at the truth about the world's ways is the very essence of scientific enterprise. (3) Abandoning of "the pursuit of truth" as a regulative ideal would hamstring from the very outset the scientific project of rational inquiry into nature. (4) The problematic nature of a "rigorous empiricism" that proscribes scientific theorizing about an observation-transcending reality must be recognized. (5) In foresaking realism, we would lose any prospect of developing a naturalistic account of why the phenomena are as they are. And this is too great a price to pay. A weighty argument against a sceptical instrumentalism is that it immediately blocks any prospect of explaining why the phenomena are as they are – an explanation that must, in the nature of things, itself proceed in ultimately non-phenomenal terms.

1. AGAINST INSTRUMENTALISM: THE DESCRIPTIVE PURPORT OF SCIENCE

Should the mutability of natural science and the fact that *our* science, as it stands, cannot be claimed to depict reality correctly be taken to mean that science has the status of a merely practical device – an instrumentality for prediction and control devoid of any actually descriptive efficacy? Does natural science perhaps furnish no depiction of nature at all, but merely provide guidance to action? Such a stance is embodied in the traditional doctrine of instrumentalism, which takes roughly the following position:

Science has no descriptive or existential import at all. It is simply an instrumentality for calculating what observational consequences will ensue (or will probably ensue) if certain things are done (or left undone) – above all, what results will be obtained when certain measurements are taken. Its theories are no more than devices for generating reliable predictions and guiding effective control. Thus science is a "black box", as it were, into which we put some information (data or assumptions) in order to get out predictions about events or instructions for modes of intervention. But the propositions that figure in the contents of this black box must be construed as devoid of any descriptive content – any claims to characterizing the nature of the world. As an instrumentality of prediction and control, science is wholly free of commitment that certain sorts of things really exist and

33

actually have such-and-such a nature. The theories adopted by science are not to be construed as assertoric propositions. The question of the truth of theories or of the existence of the things they envision simply does not arise. Theories are mere rules for drawing inferences to actual or possible observations: successful and fruitful as guides to prediction and control. The theories of natural science thus do not deal with objectively real *things* and their *modus operandi* at all, but merely provide observation-coordinative *rules* which, being rules, may be more or less useful, but will not be true or false.

As the instrumentalist sees it, we must abandon the traditional view of science as a venture in securing information about the *modus operandi* of a nature that underlies observable phenomena and provides for them through causal mechanisms. Our scientific theories are no more than a practical device – a set of rules that facilitate effective interventions and verifiable predictions, mere tools for guiding future observational expectations in the light of past experience.[1] As one recent writer put it, to accept a scientific theory is not to accept it as *descriptively true*, but rather merely as *empirically adequate*; it is not to endorse the theory's substantive declarations, but merely its observational predictions.[2]

Instrumentalism refuses to understand the theoretical claims of science at their descriptive face value, but insists on subjecting these claims to a reinterpretation that confines their bearing to the phenomenological/ observational realm. It maintains that theorizing science should not be conceived as simply involved in the traditional venture of description, assertion-as-true, and causal explanation. Instrumentalism regards science as a strictly practical endeavor that does no more than guide our expectations and canalize our interventions in the world.

Kantian terminology is useful here. Kant distinguished between objective *reality* and objective *validity*, holding that space, for example, is not objectively *real* at all, but is objectively valid or "empirically real" in that we can relate it to the phenomenal objects of our experience. It must characterize all cognitively meaningful *phenomena* but has no self-subsisting *ontological* standing at all (CPuR, B44). Much the same sort of distinction is at issue in an instrumentalist view of the "laws of nature" in our scientific theories. In particular, the theoretical entities of scientific theorizing do not have a mind-independent ontological status – are not objectively "real things" at all (not *entia per se*). They only have an "empirical reality" in facilitating explanation, prediction, and control. Like the equator, they "lie in the eyes of the beholder" as serviceable fictional items devised by their users (they are *entia per aliud*). And to have a legitimate role in an appropriately devised cognitive framework is one thing; actually to exist is another. Never the twain need meet.

Accordingly, instrumentalism insists that it is necessary to reorient the goal structure of science away from its traditional teleology of answering our questions about the causal *modus operandi* of nature. The instrumentalist prohibits us from taking at face value those claims that natural science gears to unobservable entities. The whole issue of the descriptive truth of theories simply drops away at this level. Natural science becomes devoid of any *ontological* implications. It coordinates overt phenomena and does not trade in covert "realities". "Science without ontology" should be our maxim.[3] Instrumentalism is a sort of deconstructionism of natural science. (It tries to do for *theoretical* entities what Bertrand Russell tried to do for *fictional* entities – to reinterpret talk that is *ostensibly* about entities in terms that bear no ontological weight at all.)

The instrumentalist wants to "play safe" with respect to ontological commitment. Waving aloft the banner of William of Ockham, he does not want to multiply entities beyond absolute necessity, and so he strives to be rid of unobservable entities – electrons, genes, electromagnetic fields, and the like. In his view, those theoretical statements of science that purport to characterize the make-up of such unobservables have no real existential and descriptive import as such. For instrumentalists, talk about unobservables is simply an oblique way of describing the behavior of observables. Even as "the equator" is no more than a useful fiction that enables us to find our way about in the world, so those supposed "theoretical entities" of natural science are purely fictional devices that are useful for purposes of prediction and application. On this approach, science is not a description of reality but a useful fiction that produces results.

Such, then, is the position of instrumentalism. But why should one take this stance? For the issue we face is not: "Is instrumentalism a *possible* doctrine?" Rather, it is: "Is it an *attractive* position?" It is not the *availability* of the doctrine but its *appropriateness* that is in question. What can be said for instrumentalism in this regard?

The instrumentalist doctrine clearly has its problems. Our scientific theories can be viewed as constituting a contentless black box or a mere computing device that provides for calculations that mediate between observational data-inputs and observational data-outputs – a mere instrument of convenience devoid of any descriptive import. But this is hardly a satisfactory position if we take the traditional view of science as an instrumentality for answering our questions about the world.

After all, we want to know "What is really going on in the world?" The very reason for being of science is *information* about the "external world" – securing answers to our questions about how things stand in nature in terms of description, classification and explanation. Admittedly, in developing science, we are very much interested in mastery over phenomena through prediction and control, but that is only a part of it. The cognitive impetus of securing answers to our questions about the world is crucial. The object of the enterprise is to provide answers to our descriptive and explanatory questions about the things and processes of this world and their modes of operation. It should enable us to understand and predict the ways of trees and people and planets by allowing us to account for the behavior of things in causal terms, answering along the lines of *why* water freezes or barometers fall. And here instrumentalism leaves us wholly in the lurch.

2. REALISM AND THE AIM OF SCIENCE

The reason why *Homo sapiens* instituted the scientific enterprise in the first place was to secure information about how things work in the world. In the cognitive setting of the equation that defines the descent of a freely falling body ($s = 1/2 \, gt^2$) there is an implicit group of explanatory stipulations of an emphatically existential and descriptive import.

- There is such a force as gravity deployed over a gravitational field engendered by material objects.
- This force is characterized by a constant quantity g specifying the "acceleration due to gravity".

And this situation is typical. The statements of natural science are generally made and received with descriptive and informative intent. An interpretative context of description links those equations of natural science with objective reality – at any rate as regards the matter of aim and aspiration. This is what the scientist (generally) hopes and strives for – and it is with the expectation of this goal that his statements are received by the laity. The prime aim and object of the scientific venture is to provide us with reliable and accurate descriptive information about the true make-up of a reality that is not of our making and where existence and nature are wholly independent of our cognitive endeavors.

Admittedly, the contentions of our science as they stand here and now may not be – nay presumably are not – actually adequate to reality itself. But we would abandon all descriptively informative aspirations if we failed to acknowledge their *purporting* (although perhaps failing) to depict matters as they stand in the real world. The intention to describe the world is a crucial aspect of the goal-structure of science – the very reason for the being of the venture (however far our actual performance may fall short of its realization). We introduce these "theoretical entities" in the first place in order to answer questions about the constitution and operation of the physical furnishings of nature.

A seemingly knock-down argument for scientific realism takes the following form:

(1)   Whatever is a physical part of what exists itself exists.
(2)   The world-as-a-whole (the physical universe) exists.
(3)   Those theoretical entities of science (quarks, electrons, black holes, etc.) are constituent parts of the physical universe.
(4)   Therefore, the theoretical entities of science actually exist.

But this argument is no better than a near miss because premiss (3) is not quite right as it stands. For this premiss to be tenable, the qualification "held to be" must be inserted: the theoretical entities of our science are only *held to be* parts of nature. But the argument does make transparently clear the realistic *purport* or intent of natural science. We develop the enterprise in order to give an account of what there is in the world.

From the very origin of natural science, it has been the aim of the enterprise to explain the obscure and unobservable in terms of that which we can investigate by the use of our senses.

[The school of Aristotle] in discussing the origins of things and the constitution of the whole universe, established many facts not only by plausible argumentation but even by the demonstrative mathematical reasoning, and for the knowledge of matters beyond the reach of observation (*ad rerum occultarum cognitionem*) they developed a good deal of material regarding matters that themselves can be investigated.[4]

The characterizing mandate of natural science is to furnish information about how things work in the world – about "what makes nature tick". In abandoning realism we would turn our back on the definitive descriptive and explanatory aims and tasks of the scientific enterprise. In attempting answers to our questions about how things stand in the world, science offers (or at any rate, both *endeavors* and *purports* to offer) information about the world. The extent to which science succeeds in this mission is, of course, disputable. (And no doubt in this discussion the issue of success in prediction and control will have to play a central role.[5]) But this does not alter the fact that science both endeavors and purports to provide realistically authentic descriptions of what the world is actually like.

The theory of sub-atomic matter is unquestionably "a mere theory", but it could not help us to explain those all too real atomic explosions if it is not a theory about real substances. If I hypothesize a robber to account for the missing jewelry, it is not a hypothetical robber that I envision but a perfectly real one. Similarly, if I theorize an alpha particle to account for that photographic track, it is a perfectly real physical item I hypothesize and not a hypothetical one. Only real objects can produce real effects. There exist no "hypothetical" or "theoretical entities" at all, only *entities* – and hypotheses and theories about them which may be right or wrong, well-founded or ill-founded. To re-emphasize, the "theoretical entities" of science are introduced not for their own interest but for a utilitarian mission, to furnish the materials of causal explanation for the real comportment of real things. And they cannot accomplish that job satisfactorily without being seen as real objects that form part of the physical furnishings of the world. After all, we develop science in order to tell the world's causal stories – to give a causal accounting for how it is that the things that happen in the world happen. If we did not (think that we) need those theoretical entities to tell that causal story, they would not be there.

Science makes assertions all right, but *guarded* assertions. When we look to *what* science declares, to the content and substance of its declarations, we see that these declarations purport to describe the world as it really is. Of course, when we look to *how* science makes its declarations and note the tentativity and provisionality with which they are offered and accepted, we recognize that this realism is of a guarded sort that is not prepared to claim flatly that this is how matters actually stand in the real world. Despite a commitment to realism at the semantic level of assertion-content, there is no longer a commitment to realism at the epistemological level of assertoric commitment. Realism prevails with respect to the *language* of science (i.e. the content of its declarations) even after it is abandoned with respect to the *status* of science (i.e. the ultimate tenability or correctness of these assertions). Thus our inability to claim that natural science as we understand it depicts reality correctly must not be taken to mean that science is a merely practical device – a mere instrument for prediction and control that has no bearing on describing "the nature of things". What science says is descriptively committal in making claims regarding "the real world", but the tone of voice in which it proffers these claims always is (or should be) provisional and tentative.

An instrumentalist will, no doubt, try to make capital of the circumstance that the realist is not actually in a position to insist that those purportedly descriptive theories actually describe matters *correctly*. But here the realist has a convenient reply:

So what else is new? Of course, those scientific theories I endorse are endorsed tentatively. I realize full well that they are no more than provisional estimates of the truth. But this is a matter not of their assertoric content (which is reality purporting), (but of their epistemic status (which is tentative and defeasible). We must not confuse the *substance* of our assertions with their *evidential standing*.

We must, of course, recognize that there is a decisive difference between what science *accomplishes* and what it *endeavors* to do. And it is thus useful to draw a clear distinction between a *realism of intent* and a *realism of achievement*. Scientific realism skates along a thin border between patent falsity and triviality. Viewed as the doctrine that science *indeed describes* reality, it is doubtless untenable, but viewed as the doctrine that science *seeks to describe* reality, it is virtually a truism. We are certainly not in a position to claim that science as we have it achieves a characterization of reality. In *intent* or *aspiration*, however, science is unabashedly realistic: its *aim* is unquestionably to answer our questions about the world *correctly* and to describe the world "as it actually is". The "real truth" – authentic truth about reality – represents a conception to which we stand committed throughout the whole project of rational inquiry because truth affords its aim, though not its actual achievement. The orientation of science is factual and objective: in aim and aspiration it is concerned with establishing the *true* facts about the *real* world, however much it may fall short of attaining this goal.

The fact thus remains that its concern to resolve questions about the real world is the *raison d'être* of the scientific project. Of course, we have no advance guarantee of success in this venture, and may well in the end have to recognize our limits and limitations in this regard. But this consideration affords no reason to abstain from doing the very best we can at providing full-fledged *descriptions* of the world. The theories of physics purport to describe the actual operation of real entities – those Nobel prizes awarded for discovering the electron, the neutron, the pi meson, the anti-proton, etc., were intended to recognize an enlargement of our understanding of nature, not to reward the contriving of plausible fictions or devising of clever ways for coordinating observations. However gravely science may fall short in performance, nevertheless in aspiration and endeavor it is unequivocally committed to the project of depicting "the real world", for in this way alone could it discharge its constituting mandate of answering our questions as to how things work in the world. The world-picture of natural science is – at best – a tentative or aspiring depiction of nature.

To accept scientific fallibilism does *not* mean that one must give up on the idea that scientific inquiry is a matter of the pursuit of truth regarding the workings of nature. Fallibilism simply means that we must make those descriptive and ontological claims of science in a somewhat tentative and provisional tone of voice. We must distinguish between the *mission* of the scientific project and its actual *achievement*. In intent, science unquestionably seeks to describe objective reality. But the circumstances in which we labor preclude our claiming that science actually achieves this aim. Yet while we must accept this fallibilistic view of science, the fact remains that the *aim* of getting at the truth about the world's ways represents the very essence of the enterprise.

Commitment to a realism of intent is inherent in science because of the genesis of its questions. The ultimate basis of the factually descriptive status of science lies in just this continuity of the issue of science with those of "prescientific" everyday life. We begin at the prescientific level of the archetypal realities of our prosaic everyday-life experience. The very reason for the being of our scientific paraphernalia is to resolve our questions about this real world of our everyday-life experience. Given that the teleology of the scientific enterprise roots in our commitment to the "real world" that provides the stage of our being and action, we are also committed *within its framework* to take the realistic view of its mechanisms.

Instrumentalism puts the cart before the horse. As far as the working scientist is concerned, scientific theories do not exist for the sake of prediction and control, but the other way round – prediction and control are of interest because they serve to monitor the adequacy of our theorizing about objective reality. Accommodation of the phenomena – "empirical adequacy" as it were – is *not* the be-all and end-all of scientific theorizing; it is merely a part of the test criteria for the adequacy of this theorizing.

Instrumentalism thus draws an incorrect conclusion from the undoubted fact of the fallibility and corrigibility of science.[6] This corrigibility does *not* mean that one should not make existential and descriptive claims about "the real world" in the context of science. It merely means that one must make them provisionally, talking in the hypothetical mode: "*If* 'our science' is correct, *then* electrons exist and have such-and-such features," etc. We must acknowledge that our science is ontologically committed in its descriptive and explanatory mission – in its intent or endeavor – though doubtless imperfect in its execution of this mandate.

3. THE PURSUIT OF TRUTH

Some theorists are tempted by the following sort of argumentation:

Neither can we claim to have attained the definitive truth in scientific matters, nor can we even say that we are approaching it more and more closely. It follows then, that there simply is no such thing as "the real truth" in this domain. This whole absolutistic notion should simply be abandoned.

It must be said emphatically that the fallibilistic tenor of the present discussion does *not* underwrite such an all-out sceptical abandonment of "the pursuit of truth".

To be sure, some of the most prominent philosophers of science of the age have given up on truth. Rudolf Carnap teaches that theorizing science should never make flat assertions but only statements of probability.[7] Again, Karl Popper has argued long and hard that we must abstain from staking claims to truth in the sciences; that scientists should never *believe* the theses they devise but view them as mere conjectures, which they must try (and even hope) to falsify and must never regard as claims of substantive fact.[8] Thomas Kuhn's reading of the history of science leads him to reject any claim that science presents us with the truth of things.[9] Why, then, not accept this verdict and follow the sceptical path in dropping all reference to "the pursuit of truth" as regards the aim of science?

sceptical path in dropping all reference to "the pursuit of truth" as regards the aim of science?

The answer is straightforward. It is manifestly the *intent* of science to declare the real capital-T Truth about things. Without this commitment to the truth we would lose our hold on the teleology of the aims that define the very nature of the enterprise of inquiry. The characterizing *telos* of science, after all, is the discovery of facts – the providing of presumptively true answers to our questions about what goes on in the world and why things go on as they do. Resolving such issues calls for espousing – and rightly espousing – various theses about it. In holding that scientific inquiry yields information about the world, one is constrained to hold that it entitles us to *accept* certain factual theses – with "acceptance", of course, to be understood as at least a tentative endorsement of the truth. Any view of scientific claims as *information*-providing must proceed on an acceptance-model of rational inquiry into "the truth" of things. Our attempts at descriptive information may misfire; we may well, of course, not actually succeed in finding "the real truth". But unless we are prepared to take a committal stance towards what we do find – unless we are prepared (at least tentatively) to *claim* truth for our findings and so to accept them (at least provisionally) as asserting what is actually the case – we must simply abandon an information-oriented cognitive stance toward the world. (It would not make sense to think of scientific inquiry as a project in truth-estimation if there were no truth to be estimated).

We have undoubtedly learned in the school of bitter experience that there is no alternative to presuming that our science as it currently stands does not achieve the definitive Truth. It is one thing to speak of "getting at the truth" in the language of aspiration and quite another to speak of it in the language of achievement. Truth *estimation* must be differentiated from truth *presentation*. The whole crucial contrast between the "real truths" of the perfected cognitive condition of things and merely purported or ostensible truths of the cognitive state-of-the-art would come apart if we abandoned our (regulative) commitment to the view that there is such a thing as the "real truth" and a mind-independent reality that determines it as such.[10] If science was not in its aim and aspiration an *attempt* to get at the real truth of things (an attempt that is, admittedly, imperfect and, as best we can tell, generally ends in failure), then the entire project of providing *information* about the world – of answering our questions about how things actually stand in extra-phenomenal nature – would become altogether unworkable. To abandon truth is to abandon the whole project of inquiry. A sceptical rejection of "the pursuit of truth" as a regulative ideal would abort the scientific project of rational inquiry into nature from the very outset.

Science does no more than endeavor to provide us with the very best answers currently available to our questions. The fact that these answers are probably incorrect and certainly incomplete is beside the point. The crux is that we have questions and want answers. The answers that science gives us are the best that can possibly be had at this time of day and thus represent the most that can currently be asked for. To abandon the whole information-seeking venture because much of the information it provides is misinformation is to overreact adversely. The sensible thing, in inquiry as elsewhere, is to do the best we can, and to settle for the best we can get. There is no reason why we should abandon scientific realism at the level of intent and aspiration.

In abandoning a dedication to the real truth, we would be reduced to talking only of what we *think* to be so. The contrast with "what actually is so" – the "real truth" of things – would no longer be available. We would now only be in a position to contrast our *putative* truths with those of others, but could no longer operate with the classical distinction between the putative and the actual, between what we think to be so and what actually is so. And at this point, the idea of *inquiry* – aimed as it is at increasing our grasp of the truth of things – would also be abandoned. It would be senseless to investigate something whose very existence is rejected, and to endeavor to estimate something that is not there – or at any rate presupposed or assumed or postulated to be there. In abandoning truth – in refraining from assuming or postulating that there is indeed such a thing as "the real truth of the matter" in regard to how things work in the world – we would no longer be able to conceptualize the project of scientific inquiry as we standardly do.

Such a substantiation of realism is, in effect, a transcendental argument for realism from the very possibility of science as it is standardly conceived. For given that science is by nature a project of explaining the world's occurrences – so that the very aim or aspiration of science involves a true characterization of nature's operations – one could straightforwardly say that a realism of sorts (i.e. a realism of aspiration) inheres in the very possibility of science. To be sure, "possibility of science" here means "possibility of fully achieving the aims of science" and not "possibility of securing the sort of science we actually have". Its thrust is aspirational rather than actualistic; it looks to truth attainment rather than to truth estimation. And so this sort of argument is not quite as powerful a defense of realism as is sometimes believed, seeing that the realism it secures is one of aspiration rather than one of achievement.[11]

4. ANTI-REALISM AND "RIGOROUS EMPIRICISM"

Contemporary instrumentalists take the stance that we, as good empiricists, must stick to experience and accordingly curtail our demands on scientific theorizing by limiting our understanding of what scientific theories do to the coordination of observables. They issue a stark injunction:

Do not run cognitive risks! Either avoid extending your theories beyond the reach of the phenomena (the actual or possible data of observation) or – should you persist in stretching your theoretical claims beyond this point – do not construe these theories as descriptive truths about the real world.

In proscribing unobservables, the instrumentalistic empiricist insists on maintaining an interpretation of scientific theorizing that disconnects it from any descriptive claims that transcend the observational level. As one recent expositor put it:

To be an empiricist is to withhold belief in anything that goes beyond the actual, observable phenomena [but then why not just *observed* phenomena?] and to recognize no objective modality in nature. . . . It [empiricism] must involve throughout a resolute rejection of the demand for an explanation of the regularities in the observable course of nature by means of truths concerning a reality beyond what is actual and observable. . . .[12]

On this telling, we are enjoined to forego any attempt to *explain* the phenomena in non-phenomenal terms, since – on *such* a construction of empiricism – recourse to anything nonobservable is denied us.

Instrumentalistic empiricism is risk averse. Like scepticism ("do not accept anything"), it is a policy of "safety first" rooted in a fear of mistakes ("do not accept anything that goes beyond the phenomenal data"). The extreme, all-out empiricist will take no cognitive chances.

He strives to avert mistakes at all costs. But like the sceptic, he pays a price for the comfort of safety and security. For if we want information – if we deem ignorance no less a negativity than error – we must be prepared to "take the gamble" of answering our questions in ways that run some risk of error. (The sceptic and empiricist can only get round this point by arguing that there is no information to be had – though how they can pre-establish this short of a fair trial is a tricky question.)

The appropriate stance here was indicated with trenchant cogency by Charles Sanders Peirce: "The first question, then, which I have to ask: Supposing such a thing to be true, which is the kind of proof which I ought to demand to satisfy me of its truth."[13]

A general epistemic policy which would make it impossible, as a matter of principle, for us to discover *something which may (for all we know) responsibly be supposed actually to be the case* is clearly irrational. And the proscription of accepting it is obviously such a policy – one which aborts the whole project of inquiry into transphenomenal reality at the very outset, without according it the benefit of a fair trial.

Two fundamentally different sorts of misfortunes are possible in situations where risks are run and chances taken:

(1) *Misfortunes of the first kind*: We decline to "take the chance" and avoid running the risk at issue, but things turn out favorably after all, and we "lose out on the gamble".

(2) *Misfortunes of the second kind*: We do "take the chance" and run the risk at issue, but things go wrong, and we "lose the gamble".

If we are risk-seekers, we will incur few misfortunes of the first kind, but – things being what they are – relatively many of the second kind will befall us. Conversely, if we are risk-avoiders, we shall suffer few misfortunes of the second kind, but – things being what they are – we shall encounter relatively many of the first. The overall situation is depicted in Figure 1, which sets the stage for deliberations about the rationality of risk.

Obviously, the sensible thing to do is to adopt a policy that minimizes misfortunes *overall*. Rationality generally calls for the "middle-of-the-road" policy of risk calculation – of acting as best we can to balance the dangers and opportunities. It makes sense to adopt the Aristotelian

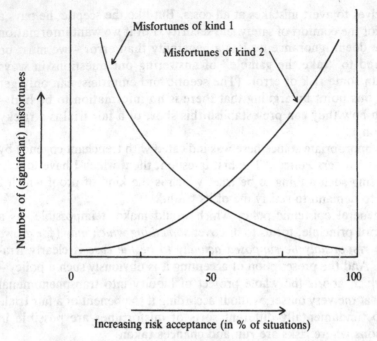

Fig. 1   Risk acceptance and misfortunes.

"golden mean" between the extremes of risk-avoidance and risk-seeking. The path of reason calls for realistic calculation and prudent management. Its line is, "Neither avoid nor court risks, but manage them sensibly in the search for an overall minimization of misfortunes."

The empiricistic instrumentalist will not risk making any claims: the prospect of falling into mistake epitomizes his idea of catastrophe. But, in cognition, as elsewhere, "nothing ventured, nothing gained".

The obtaining of *information about the world* is, after all, the aim and purpose that governs the entire cognitive venture. To be sure, when we set out to reach this goal, we may well discover in the end that, try as we will, success in reaching it is beyond our inadequate means. But we shall certainly not reach the goal if we do not set out on the journey at all, which is exactly what the blanket proscription of acceptance amounts to. The rigorous empiricist writes off at the very outset the prospect of explaining the phenomena in extraphenomenal terms – a prospect whose rejection would only be rationally defensible at the very end. He is fundamentally irrational in taking a stance that commits Peirce's cardinal sin of blocking the path of inquiry.

On the basis of our experience in this world we (1) have questions, and (2) have formative conceptions as to the sorts of answers to these questions that can qualify as (potentially) satisfactory. And this is what renders those "Do not ask!" prohibitions inherent in the phenomenogical restrictions that characterize instrumentalism so profoundly dissatisfying. We are, to be sure, prepared to give up our questions or our preconditions for a "satisfactory response". However, we would be willing to do this only *in extremis* – under the pressure of a dire last-ditch necessity that leaves no viable alternatives open. But in the case of our efforts at scientific inquiry, this necessity just is not there. The tentative and corrigible nature of our theoretical scientific claims is clearly something we can learn to "live with". Their descriptive vacuity – their lack of pertinence to anything transphenomenally objective – is something else again.

5. THE PRICE OF ABANDONING REALISM

The instrumentalists' limitation of strictly construed scientific discourse to the realm of phenomena represents an unacceptably stringent curtailment. In theory one can be increasingly liberal in one's ontology by recognizing the following three categories:

(1)   Phenomena (observations and measurements) and the relations among them.
(2)   Things and processes that cause the phenomena.
(3)   Principles of operation ("laws") that govern the things and processes that cause the phenomena.

The radical empiricist says: "Avoid complications and potential problems; simply stop at 1, refusing to go beyond the phenomena themselves." As one recent neo-Berkeleyan theorist puts it:

That the observable phenomena [have the character they do] . . . . is merely a brute fact, and may or may not have an explanation in terms of non-observable fact "behind the phenomena". . . . [14]

For the thrust of traditional empiricism from Berkeley and Hume onwards runs roughly thus:

Don't trouble us with conjectural substances or principles lying outside the reach of observation. Stick strictly to the realm of experience. Don't encumber the world with imperceptible substances or with causal laws that reach beyond the level of observation.

Contemporary empiricists continue to use exactly this same classical strategy. Does it enable them to consolidate a victory in the philosophy of science?

In addressing this issue it helps to go back to square one and ask: What is empiricism all about? Philosophers of science seem widely agreed that it is a good thing to be an empiricist. But there is no consensus as to exactly what sort of commitment this "empiricism" involves. There is a considerable variety of substantially different positions:

(1)   Keep theories to mere summaries of the actual data of observation. Stick to the observed phenomena and do not go beyond them – or at any rate no further than the sort of interpolation at issue with Hume's missing shade of blue. Theories should simply *classify* the observed phenomena.

(2)   By all means let your theories go beyond the *actual* data of observation, but only as far as *possible* data. Theories and data must be subject-matter coordinated – so, in particular, theories must introduce no sorts of entities not present in the data. Theories should simply be general rules that coordinate phenomena in lawful patterns.

(3)   Let there be no holds barred on the contents of your theorizing. Project your theories as far beyond the data as you like. But just do not *believe* them. Do not view them as literally true stories about the world's furnishings and their modes of comportment. Always remember that they are mere useful fictions to enable you to operate amidst the observational data in point of prediction and control. Theories must not support extraphenomenal conclusions: they must simply accommodate or save the phenomena.

(4)   Feel free to believe your theories. But keep them sensitive to the observational data. Only theorize in such a way that if the phenomena were sufficiently different, the theories would be correspondingly different.

Here (1) is relatively restrictive and (4) relatively liberal. Certainly, the empiricist formula "There are no unobservable entities" is a contention which, in one particular sort of construction, merits emphatic endorsement. For *no physical entities are, in principle, totally unobservable, totally undetectable in nature*, if we construe "detectability" broadly as involving whatever we can discern in some way or other in which theory will certainly play a role. The key question is: Just how far down this list of such a spectrum of increasingly weak empiricisms is one to go?

All of these four positions are "defensible" in some mode or other; all are serious candidates for consideration – there is something to be said for each. The choice is ultimately one of epistemic risk-tolerance. It pivots on the question: How far are we willing to go in chancing the possibility of error in the interests of getting a fuller and richer account of the world's nature?

With each added step of liberalization, with each relaxation of restrictions on legitimate theorizing, we run greater cognitive risks. We confront a more complex ontology, a more elaborate set of ontic commitments, a more extensive transcendence of evidence at hand. But at the same time we secure a richer and fuller account of the world and its workings, a more far-reaching and comprehensive set of answers to our questions. Our understanding is increasingly enhanced. And just this consideration speaks out against an overly rigid construction of empiricism. We can agree with empiricists that experiential knowledge *begins* with phenomena without conceding that it must also *end* there.

Accordingly, the realist is able to tell a more rounded causal story. He does not have to respond at virtually all points with "Do not ask!" and proscribe all sorts of questions regarding the structure and causal operation of physical things as transgressing the limits of evidential propriety.

The instrumentalistic antirealist who maintains that science aims not at descriptive truth, but merely at empirical adequacy, reproves the theorizing scientist roughly as per in the following passage:

Don't you scientist fellows go getting pretentious on us. It is not for you to tell us what is real in the world. (That is something for the theologian, the philosopher, or the plain man of ordinary good common sense, or whatever, but at any rate not for you theoretical fellows.) So do not come to us with immodest talk about the reality of things like genes or electrons or quarks or gravitational fields or radiation belts. Bear in mind that all this sort of thing is just shorthand for talk about phenomenal regularities. Just toe the line of a rigorous empiricism and leave off all that reality talk. What is real in the world is none of your business. Your concern is simply with hypotheses that "save the phenomena" of observation. So just stick to talk about "empirical adequacy to observation" and leave consideration of what is real in the world to others.

What we have here is little news to anyone familiar with the trials and tribulations of the late Galileo Galilei. For what Cardinal Bellarmine urged on Galileo in 1616 was to view his astronomy in exactly this way – as a strictly computational enterprise for predicting apparent planetary positions without any claims regarding the physical mechanisms involved in planetary motion.[15] Bellarmine wanted to return the state of play in natural science to exactly the point at which it had been left by Osiander's preface to *The Revolutions of the Heavenly Spheres* by Nicholas Copernicus:

It is the duty of an astronomer to compose the history of the celestial motions through careful and skillful observation. Then turning to the causes of these motions or hypotheses about them, he must conceive and devise, since he cannot in any way attain to the true causes, such hypotheses as, being assumed, enable the motions to be calculated correctly from the principles of geometry, for the future as well as the past. The present author, Copernicus, has performed both these duties excellently. For these hypotheses need not be true nor even probable; if they provide a calculus consistent with the observations that alone is sufficient.[16]

Exactly this insistence that scientific theories are mere utilitarian hypotheses that must not be seen as literally true or even probable but as devoid of any descriptive, ontological claims – simply as coordinating the phenomena – is at work in contemporary instrumentalism. And in the eyes of realists this insistence constitutes its decisive disability.

It is a telling argument against a sceptical instrumentalism that destroys any hope of answering our questions about why the phenomena are as they are. If we are flatly prohibited from invoking real entities or processes that "lie behind the phenomena" as their (non-phenomenal) causes, then at one stroke we lose any and all chance of developing a strictly naturalistic account of the causal genesis of our phenomenal experience. When the range of our acceptable claims is confined to the phenomenal sphere, we are deprived of the mechanisms through which alone we can construct a picture of ourselves as one item of physical reality among others, constructing our picture of nature in consequence of various sorts of interactions with it. We lose the prospect of developing a thorough-going naturalism based on the causal interaction view of the genesis of experience. When we are limited to the sphere of phenomena alone, we can no longer account for those phenomena through explanation in external terms. We are constrained to accept observation – and phenomenal experience in general – as a "gift horse" we are enjoined from examining too closely.

To give up on scientific realism by adopting a strictly instrumentalistic reconstruction of scientific theories is to take a step that exacts a great price in bafflement and incomprehension. For if our sense-provided experiences are *not* the causal product of the operation of independent agencies, then we cannot explain:

(1)   Why these phenomenal regularities should pervade our own (personal) experiences.
(2)   Why these regularities of ours should be systematically connected with the experiences of others.
(3)   Why we have no control over the phenomena through thought alone.
(4)   Why we can achieve (limited) operational control over nature's phenomena.

In abandoning the idea of a non-phenomenal order from which the phenomena themselves emerge through causal processes, we destroy any prospect of explaining why the phenomena are as they are – why our observations are as we find them to be. In adopting instrumentalism, we purchase the ontological economy of a doctrine that rejects theoretical entities at the cost of lost understanding. An avalanche of unanswered questions now overwhelms us. Not only are we given no answer to our questions of why the phenomena are as they are – no explanation of the phenomena in non-phenomenal terms – but we are precluded from even *asking* such questions. On the instrumentalists' approach we simply confront brute facts throughout this range: there is no prospect of explanation through the operation of underlying mechanisms. The quest for explanatory understanding has come to the end of its tether.

Realism, by contrast, can develop an explanation of phenomenal regularities – a "picture of nature" that makes the whole thing intelligible. It offers us a naturalistic approach to experience as causally emergent from a sub-experiential order of things and processes. It has a perfectly sensible tale to tell – a naturalistic account of man as one physical system among others, interacting with the rest of physical reality in various ways that give causal rise to the phenomena of conscious observation. No doubt this is a story which as yet has many gaps, but, in principle, these can be filled in as we go along with the further development of science. However, the science that can do the job must clearly be one that is construed realistically.

A consistent anti-realist has little alternative but to join the group of recent writers such as Bas van Fraassen and Richard Rorty who – on analogy with the fox and the grapes of Aesop's story – proceed to dismiss the impetus of explanation and the quest for intelligibility as undesirable and outmoded. In the cost-benefit analysis of the cognitive domain this is surely a very high price to pay – so high we would surely pay it only if absolutely necessary. And this is certainly not so.

At the close of his 1961 survey of the competing claims of realism and instrumentalism, Ernest Nagel concluded that "the opposition between these views is a conflict over preferred modes of speech".[17] If we were willing to substitute "thought" for the somewhat tendentious "speech" ("it is only a *façon de parler!*"), and were willing to recognize that people's sensible preferences are generally formed with a view to what is actually *preferable*, then this position of Nagel's emerges as a not unreasonable one. Indeed, the conflict is "ideological" and normative, reflecting a disagreement in cognitive values regarding the *appropriate* mode of interpreting scientific theories. To re-emphasize, the salient question is this: how much cognitive risk will we accept to get a satisfying account of the causal background of the phenomena?

Ultimately, it is a question of value trade-offs that we face: Are we prepared to run a greater risk of error to secure the potential benefit of an enlarged understanding? The instrumentalist is a cautious empiricist who insists on risk-minimization; the realist is a more relaxed empiricist who insists on the value of understanding. It is a matter of priorities – of *safety* versus *information*, of *ontological economy* versus *cognitive advantage*, of *epistemological risk-aversion* versus the *impetus to understanding*. The ultimate issue is a question of value: Is the gain scientific understanding worth the loss of ontological economy that would result from abandoning theoretical entities?

On this perspective, the debate between instrumentalism and realism emerges in the final analysis as a controversy about values. It turns on the appraisal of the relative seriousness of ignorance and the vexation of unanswered questions on the one hand and prospective error and the embarrassment of claim retraction on the other. We can seldom achieve absolute refutations or validations in philosophy. In general, all we can ever do is to develop a sort of cost/benefit analysis, reckoning up the advantages and disadvantages of various views – their assets and liabilities – and comparing these with one another, assessing where the best option lies. In crudest terms the issue comes down to this: Which one has the worse negativity, to be uninformed or to be wrong? "You pays your money and you takes your choice."

The weakness of instrumentalism is that its sole apparent advantage is one of ontological parsimony, so that the sole apparent price we pay for its rejection is a loss in ontological economy. This brings to the forefront the issue of the extent to which ontological economy is a good thing. We confront the question: Just what does ontology economy do for us? Why should it count as a substantial advantage – indeed, a virtually decisive one?

In addressing this issue, it is useful to consider a contrast of the situation of *ontological* economy with that of an *explanatory* economy of principles. An economy of principles – of axioms in a system of mathematical physics, for example, has clear advantages. The fewer the principles, the easier their testing and confirmation. This sort of explanatory economy immediately yields significant practical advantages for the conduct of inquiry in terms of convenience and ease of functional procedure. Ontological economy has no comparable automatic advantage.

Realism's advantage, then, is that it makes possible an explanatory reduction of phenomenal regularities via laws of operation of a subphenomenal domain. The advantages of such a systemic economy of explanatory principles is something that realism affords us but that a phenomenalistic instrumentalism must by its very nature forego.

The realist who wants to explain the phenomena on the basis of an underlying realm of extraphenomenal reality has nothing against Ockham's razor principle as such. He too does not want to multiply entities beyond necessity; he too wants no more entities than he requires for the accomplishment of useful explanatory work. He only asks for what is required to do an eminently useful job.

The trouble with a phenomenalistic instrumentalism is that it treats ontological economy as a prime good – one that is worth purchasing even at a considerable cost in terms of a wider economy that embraces the economy of principles as well. Ontological parsimony is all very well when it comes free of charge – or at any rate cheap. But it may well be bought at too dear a price if it requires us to pay substantial costs in terms of the intellectual satisfactions of explanation and understanding. As the realist sees it, phenomenalistic instrumentalism purchases its ontological economy at the cost of significant systemic complications in circumstances where there is simply no good reason to think that this price is worth it. No good reason, that is, unless we regard ontological parsimony as a substantial good in its own right which, as such, is well worth the price of systemic complication and explanatory impoverishment.

In this instance, all one can do is to ask just why it should be that ontological economy is so great a good. And when this question is addressed to the instrumentalistic/phenomenalist tradition it is met with stony silence. To all appearances, we have entered a domain where right-minded people ask no questions, a region where true believers somehow see that this is how it has got to be!

A rigoristically instrumentalistic empiricism will hear nothing of what does not meet the eye – it wants nothing of what is hidden beyond the horizon of phenomena. But because of this very stance, one telling argument that can be developed against a sceptical instrumentalism is that it constitutes an unjustifiable impediment to our impetus towards the explanation of why the phenomena are as they are – an explanation that must, in the nature of things, itself proceed in ultimately non-phenomenal terms.

On the crucial question of how the phenomena came to be as they are, instrumentalistic phenomenalism enjoins us to an embarrassed silence. Clearly, this is not a very satisfying position.

CHAPTER FIVE

SCHOOLBOOK SCIENCE AS A BASIS FOR REALISM

SYNOPSIS. (1) The theories and theses of natural science at the technical frontier are highly vulnerable. This vulnerability of our scientific estimates inheres in their generality and precision. Ordinary-life "knowledge", on the other hand, manages to gain security by reducing its definiteness or informativeness. In consequence, however, the sort of "understanding" it provides is not altogether satisfactory. There are problems on both sides. Scientific knowledge is too vulnerable: everyday knowledge too vague. (2) To get information about reality that is in a substantial degree *both* scientifically informed *and* secure, we must fall back from frontier science to rudimentary "schoolbook science". (3) A plausible scientific realism must be based on such "softened-up" popularized science rather than authentic, technical science at the frontier of research. This compromise position represented by "schoolbook science" provides for a viable form of science-indebted realism.

1. THE SECURITY/DEFINITENESS TRADEOFF AND THE CONTRAST BETWEEN SCIENCE AND COMMON SENSE

It is of the nature of technical natural science, as its practitioners actually pursue it at the research frontier, to state how things stand exactly and how they go always and everywhere – in full generality and precise detail. Technical science foreswears any "loose talk" at the level of vague generality or analogy or approximation. It has no use for qualifiers such as "usually" or "roughly". Universality and exactness are its touchstones. This circumstance renders the claims of science vulnerable – vulnerability being the price that we pay for generality and precision. We know that none of the hard claims of present-day frontier natural science will move down the corridors of time untouched.

The contrast between the world-view of science and that of common sense repays closer scrutiny as regards this matter of vulnerability.

Increased confidence in the correctness of our estimates can always be purchased at the price of decreased accuracy. We estimate the height of the tree at *around* 25 feet. We are *quite* sure that the tree is 25±5 feet high. We are *virtually certain* that its height is 25±10 feet. But we are *completely and absolutely sure* that its height is between 1 inch and 100 yards. Of this we are "completely sure" in the sense that we deem it "absolutely certain," "certain beyond the shadow of a doubt," "as certain as we can be of anything in the world," "so sure that we would be willing to stake our life on it," and the like. With any sort of estimate there is always a characteristic trade-off relationship between the evidential *security* of the estimate on the one hand (as determinable on the basis of its probability or degree of acceptability), and its contentual *definiteness* (exactness, detail, precision, etc.) on the other.

This relationship between security and definiteness is generally governed by the limits set by a curve which roughly takes the form of an equilateral hyperbola: $s \times d = c$ (c = constant). (See Figure 1.) No doubt, with the progress of science we can increase the value of parameter c somewhat. But the fundamental trade-off relationship remains. An information-theoretic uncertainty principle prevents our obtaining the sort of information we would ideally like.[1]

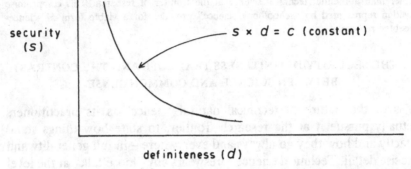

Note: The overall quality of the information provided by a claim hinges on combining its security with its definiteness. Given suitable ways of measuring security (s) and definiteness (d), the curve at issue can be supposed to be an equilateral hyperbola obtained with $s \times d$ as constant.

Fig. 1.   The trade-off between security and definiteness in information.

And so, the exactness of technical scientific claims makes them especially vulnerable, notwithstanding our most elaborate efforts at their testing and substantiation. Science declares not merely that roughly such-and-such occurs in roughly certain circumstances, but exactly what happens in exactly what circumstances. In science we always aim at the maximum of universality, precision, exactness, etc. The law-claims of science are *strict* – precise, wholly explicit, exceptionless, and unshaded. They involve no hedging, no fuzziness, no incompleteness, and no exceptions. In making the scientific assertion that "the melting point of lead is 327.7°C", we mean to assert that *all* pieces of (pure) lead will unfailingly melt at *exactly* this temperature. We certainly do not mean to assert that *most* pieces of (pure) lead will *probably* melt at *somewhere around* this temperature. (And at this point there is thus a potential problem should it turn out, for example, that there is no melting *point* at all and that what is actually at issue is the center of a statistical distribution.)

By contrast, when we assert in ordinary life that "peaches are delicious" we are asserting something like "most people will find the eating of suitably grown and duly matured peaches a relatively pleasurable experience". Such a statement has all sorts of built-in safeguards such as "more or less", "in ordinary circumstances", "by and large", "normally", "if all things are equal", and so on. They are not really laws but rules of thumb, a matter of practical lore rather than scientific rigor. And this enables them to achieve great security. For there is safety in vagueness: a factual claim can always acquire security through inexactness. Take "there are rocks in the world" or "dogs can bark". It is virtually absurd to characterize such everyday-life contentions as fallible. Their security lies in their indefiniteness or looseness – it is unrealistic and perverse to characterize such common-life claims as "defeasible". They say so little that it is "unthinkable" that contentions such as these would be overthrown.

In natural science we deliberately court risk by aiming at maximal definiteness, and thus at maximal informativeness and testability. Aristotle's view that science deals with what happens ordinarily and "in the normal course of things" has long ago been left by the wayside. The theories of modern natural science take no notice of what happens usually or normally; they seek to transact their explanatory business in terms of strict universality – in terms of what happens always and everywhere and in all circumstances. And in consequence we have no choice but to accept the vulnerability of our scientific statements relative to the operation of the security/definiteness trade-off.

This relationship is thus the basis of scientific fallibilism. The increased vulnerability and diminished security of our claims is the undetachable other side of the coin of definiteness. Science operates in the lower right-hand sector of Figure 1. Its cultivation of informativeness (definiteness of information) entails the risk of error in science; its claims are subject to great insecurity. The frontier theories of natural science have a relatively short half-life.

However, the ground rules of ordinary-life discourse are altogether different from those of natural science. Ordinary-life communication is a practically oriented endeavor geared to social interaction and the coordination of human effort in communal enterprises for the common good. The operative injunctions here are: "Aim for security, even at the price of definiteness; do not lay yourself open to the reproach of purveying mistaken information. Avoid misleading people into error. Preserve your credibility!" In ordinary life we operate at the upper left-hand side of the Figure 1 curve. The situation contrasts sharply with that of science, whose objectives are largely *theoretical*, and where the name of the game is disinterested inquiry.

Accordingly, very different probative orientations prevail in the two areas. In ordinary-life contexts, our approach is one of situational satisficing: we stop at the first level of sophistication and complexity that suffices for our immediate needs. In science, however, our approach is one of systemic maximizing: we press on towards the ideals of systemic completeness and comprehensiveness. In science we put ourselves at greater risk because we ask more of the project.

In this contrast between natural science and common sense, science wins out in point of precision and detail, but common sense wins out in point of security. At the level of scientific generality and precision we have no secure and invulnerable information, while at the level of common sense we get no detailed understanding. As the aims of the enterprises are characteristically different, our inquiries in everyday life and in science have a wholly different aspect, with the former achieving stability and security at the price of sacrificing definiteness, a price which the latter scorns to pay.

A possible misapprehension arises in this connection. A contention along the following lines is very tempting: "Science is the best, most thoroughly tested knowledge we have – the 'knowledge' of everyday life pales by comparison. The theses of science are really secure and well-established, those of everyday life casual and fragile." But in fact the very reverse is the case: our scientific theories are vulnerable and have a shortish lifespan; it is our claims at the looser level of ordinary life that are very secure and stable.

It thus appears that in managing the business of realism we should put our trust in the bank of common sense rather than in that of natural science. And yet common sense too has its problems. For its concepts are loose, fuzzy, general. Its world is populated with things such as trees, rocks, gasses and all those conceptions of a rudimentary "knowledge" that we deem too loose and sloppy for the purposes of explanatory understanding. The adequacy of this sort of knowledge in point of accuracy (rather than security) is something that science has undermined for good and all. The scientific project itself teaches us that the concepts of everyday life are too vague to characterize adequately the world as it really is – seeing that it was the unsatisfying character of the "stone-age physics" of untutored common sense that drove us to develop science in the first place.

Our ordinary, common-sensical concepts such as "tree" and "rock" are useful enough in their way, facilitating the coordination of practical activities, for example, working together to clear a field. But they lack the precision required for satisfactory explanations. And yet, in seeking this precision, science courts error; the claims of the state-of-the-art in science at any particular time doubtless need eventual revision and correction. To characterize reality adequately, we need theses that are both relatively secure and relatively precise. And yet we are caught betwixt and between. Common sense is capable of yielding secure claims, but only vague ones; science yields only insecure claims, but precise ones.

We are caught therefore between the two. What we need for realism is information about reality that is both tenable (secure) and highly informative (definite). But just this is something that the exclusion principle encapsulated in the security/definiteness trade-off renders problematic. From the angle of precision, science is in good shape, but its condition is very problematic from the angle of truth. Common sense, on the other hand, stands on firmer ground in relation to security, but is so untidy conceptually that it lacks precision. Either way, the epistemic component of realism is in difficulty. Is there a middle ground? There is indeed! It is elementary school level science-made-simple that comes to our rescue.

2. SCHOOLBOOK SCIENCE AND "SOFT" KNOWLEDGE

What we desire and require is a tenantable halfway-house between an untenable scientific realism subscribing to the mistaken supposition that "the theories of natural science correctly describe physical reality", and a description-abandoning instrumentalism subscribing to the bankrupting thesis that "the theories of natural science should be construed nondescriptively, as merely coordinating the phenomena". Exactly such a halfway house is provided by what we have characterized as "school-book science", which is able to provide a realism of the middle range – the realism of the elementary science of popularizations and school texts.

Are there atoms? Is the atomic theory true? It all depends. It is matter of just what the "atomic theory" is – just what it is to be taken to assert.

Taken at face value, "the atomic theory" that we now have is surely problematic. After all, *exactly* what are atoms like? Neither Rutherford, nor Bohr, nor Weinberg–Salam, nor the nuclear physics of today provide a definitively *correct* picture of the atom. We clearly cannot trust the science of the day to the extent of saying that we know *just exactly* what atoms are like.

All the same, atoms at large (or better, "at loose") are altogether safe. There is no danger that atoms and molecules will cease to play a role in physics and chemistry. Even though we shall doubtless have to revise our conceptions of them, the things as such will not go away.

The looser and detail-suppressive knowledge of popularized science – indebted though it is to technical science – is not itself scientific knowledge. Technical science scorns imprecision and operates at the outer limits of exactness and accuracy. It would not stoop to say the sorts of detail-ignoring things we find in science reporting or in introductory accounts for laymen. But it is just this sort of popularized "schoolbook science" that is the focus of our confident assurance that we know how things stand in the world.

One recent writer poses the following worry. Recent science abandons the luminiferous aether. What if future science did the same with the electron?

Then we will have to say electrons do not really exist. What if this keeps happening? What if *all* the theoretical entities postulated by one generation (molecules, genes, etc., as well as electrons) invariably do not exist from the standpoint of later science?[2]

But this worry is simply another instance of the "what if" thinking of traditional scepticism. ("What if life were a dream?") Such hand-wringing, abandoning electrons, molecules, and genes (which, after all, we can "observe" with contemporary technology) is hyperbolic. We can – and doubtless, will – come to think these things *very differently* from the way in which we conceptualize them today. But given the conceptual plasticity and flexibility of what is at issue with "electrons", "genes", etc., the prospect of their total disappearance is on the same level as the total disappearance of Julius Caesar. It is not that this is something *logically* impossible. It is simply that this is one of these unrealistic possibilities that need not worry us in conducting practical policies for our theoretical concerns. In succumbing to this sort of worry, we leave the issue of science behind and plunge into the generalized scepticism of far-fetched what-if "speculation".

Moreover, the argument of the preceding passage has another defect when seen as an objection to scientific realism. Admittedly, scientific entities can become dislodged – phlogiston moved aside to make room for oxygen, the "gravitational attraction" of Newton gave way to the space-warp of Einstein, radiation succumbed to fields and ceased to be a matter of waves-in-a-medium, thereby abolishing the luminiferous aether. But the disappearance of *some* "scientific entities" does not mean that they *all* disappear. On the contrary, they do not *vanish* at all; they are *replaced*: it takes one to dislodge one. And so the preceding "argument from error" is not an argument against scientific realism, but simply one against the ontological finality of science as we have it. At first people thought that man's proto-ancestor was akin to a gorilla, then the baboon came into favor, only to be displaced eventually by the chimpanzee. But such a series of changes does not support the induction that man has no ancestor at all. Similarly, we cannot, in the case of science, construe the historical sequence of changes of mind about theoretical entities as providing an argument against the reality of theoretical entities at large.

We cannot simply rely on "the present state of science" to give us an unproblematically acceptable picture of the world, because we realize that this picture is defective. The frontiers of science will move on, and with the wisdom of hindsight it will come to be seen by our successors as every bit as inadequate as we ourselves deem the science of our predecessors of 100 years ago. But our knowledge of the world nevertheless rests on scientific knowledge. It does so by taking the hard scientific information of the day and "loosening it up" by removing all that fine-grained detail. We constantly get more information of high security, albeit information that is loose and imprecise. With "the growth of our scientific knowledge" (vulnerable though it is!) there is an ever-increasing number of things we can claim with high confidence.

Certainly, our looser schoolbook science is not invulnerable – it too is not absolutely secure and definitive, but will doubtless undergo sea-changes in the course of time. The point is rather that "it is all relative" – that schoolbook science is *very secure* in comparison to technical science and *very informative* in comparison to the "stone age physics" of rational common sense.

This loosened-up "popularization" of science at issue in schoolbook science is, of course, generically "scientific" – it is indebted to technical science and undergirded by it. But it is insufficiently detailed and therefore insufficiently deep – unable to answer adequately to our demands for a detailed understanding how things work in the world. (That is just what marks it as "popular" rather than "real" science.) All the same, it gains its crucial advantage of security precisely because of its looseness. It achieves that middle ground of concurrent security and informativeness that is essential to any viable sort of information about the real. Absolutely definitive knowledge of nature being unattainable, we can and should be content to let our realism operate within the limits of the attainable represented by the middle range of the security/ definiteness curve.

3. SCHOOLBOOK SCIENCE AS A BASIS FOR REALISM

Although the claims of our technical science will require eventual correction at the level of detail and precision at which they are cast, nevertheless the overall picture that emerges from science is doubtless right in its rough outlines. An increasingly adequate picture of nature emerges not *in* technical science, but *through* technical science. We realize that we have to be prepared to revise any and every thesis of "hard" science – that none of its present day theories will survive unscathed to the year 5 000 – wholly without revision. But this is not the case with the grosser materials of schoolbook science. (Atoms of some sort will be with us from here on in.)

Schoolbook science not only claims *that* there are atoms, magnetic fields, and genes, but obviously also involves some claims about *what* they are like. Yet they are only rather rough and inexact claims. Unlike substantive science – an aggregate of detailed theories – it involves no commitment to the precise details of any particular theory whatsoever. A realism of schoolbook science opens up the clearly plausible prospect of being realistic about the theoretical entities of science (holding they are real) without being realistic about the current theories of science (holding they are actually true and give a *correct* account of reality).

Such a realism of schoolbook science is a physical realism that is neither merely a common sense realism nor a strictly scientific realism, but resides in a halfway-house represented by popularized science. It is a softer scientific realism that draws on technical science not for its details but for the rough essentials.

This approach puts us in the fortunate position of being able to hold that the objects discussed in our scientific theories have a life independent of those theories and do not stand or fall by their correctness.[3] It is precisely our acknowledgement of the looseness of our knowledge about atoms and molecules that renders the existence of such things secure. If the existence of something is subject to the formula "to be is to be (exactly) as we now deem it to be" – then its existence would stand on very shaky ground indeed.

The great advantage of a realism of schoolbook science is that it disconnects the issue of the existence of "theoretical entities" from the tenability of our particular theories, making it possible for such entities to have a life independent of our current beliefs about them. It makes it possible to say (surely rightly!) that we need not maintain the definitive truth of any of the current formal theories of natural science in order to maintain that science provides us substantial information about the workings of nature at the level of observables and unobservables alike.

What we obtain on this basis is not a full-blown "scientific realism" that claims the entities discussed in technical theoretical science exactly as this science describes them. It does not maintain that we can read off the way the world is from the statements endorsed by natural science as it stands. Rather, it is a realism that is indebted to science but does not involve reifying the theoretical entities of science in just the way currently envisioned by natural science. It is a realism that becomes available not *in* but *through* natural science as we have it, not a scientific realism, but a science-indebted realism.

The realism of schoolbook science leaves us very much with a half-full barrel. It puts us well ahead of a barebones metaphysical realism that merely maintains that there is a mind-independent reality, but it is unable to provide detailed descriptive information about this reality. For on the basis of schoolbook science, we can know a good deal about nature – and can come to know ever more about it as inquiry proceeds. All the same, what we learn at this level of "schoolbook science" is vague, imprecise, and general, rather than specific, exact, and accurate. It is gravely deficient in informativeness, giving us a picture of reality seen "through a glass, darkly" without the precision and detail of a scientifically exact idea that we would dearly love to have.

Nevertheless, as technical science develops, the schoolbook realism it pulls along in its wake enlarges as it grows both in scope and in detail. Such "improvements" never take us as far as we would like – of necessity the position we reach is always far reached from the cutting edge of precision and detail. Still, this half-full barrel is not empty. It combines a recognition of the fallibility of natural science with a robust realism that talks of truth and real existence *even when unobservables are involved* – a realism that accepts electrons and genes (answering to rough-and-ready descriptions) without categorically affirming our current scientific conceptions about such things.

This is where schoolbook science gains its utility for realism. Even as there is more to an apple or a piece of rock than our own potentially incorrect ideas about them, so there can be more to an electron or a gene than the current theories of science-as-it-stands envision. It is thus perfectly possible to be an ontological realist about "theoretical entities" without being a semantical realist (i.e. truth-endorser) with respect to the current *theories* in which they figure. We are relieved of the need to think that our theories fit their objects perfectly – those objects stand or fall by the correctness of our present ideas about them. There is (and should be) more to our "theoretical entities" than meets the eye of current theorizing, seeing that our theories presumably get it wrong, doing no better than to have "the right general idea". It is here that "schoolbook science" comes into its own and is able to provide a crucial support for realism.

A realism based on schoolbook science is an attractive position because it is able to reconcile two facts – on the one hand, we cannot claim that natural science as it stands characterizes reality correctly, but on the other hand, we cannot simply dismiss it as totally uninformative about "the way the world is". In the choice between a naive scientific realism and a sceptical scientific fallibilism, recourse to schoolbook science affords us a middle way enabling us to acknowledge the ambivalence of our attitude towards the deliverances of current science.

CHAPTER SIX

DISCONNECTING THEIR APPLICATIVE SUCCESS FROM THE TRUTH OF SCIENTIFIC THEORIES

SYNOPSIS. (1) It is often said that the successes we realize in applying the theories of science betoken their truth. But this is a very dubious proposition. (2) For their actual truth is not the *best* explanation of the successful application of our scientific theories. An *incapacity to distinguish* the currently efficacious from still superior, but not yet projected, theories affords an even better and more realistic explanation. (3) The successful application of a scientific theory is a highly ambiguous datum. (4) All we can conclude from such a successful application is *that there is* an adequate explanation of why our theory provided this success – but we cannot conclude that this explanation is provided by the theory itself and lies in the fact that it is true. (5) The gap between success and truth does not, however, give comfort to the anti-realist. The fact that our cognitive grasp of reality is imperfect constitutes no argument against realism.

1. IS SUCCESSFUL APPLICABILITY AN INDEX OF TRUTH?

Scientific realists frequently maintain that the successful applicability of scientific theories in matters of prediction and control is a sure token of their truth. And, indeed, it *seems* eminently plausible to argue as follows:

If nature were not actually and in real fact much as natural science claims it to be, then the applicative and particularly the predictive success of science would be a miracle. It would be like someone ignorant of the right number for dialing a friend on the telephone succeeding in telling us how to do it. In both cases alike, if you do not have it right, it is virtually certain that you will not make the right connections.[1]

A fallibilist who does not accept that science as we currently have it is substantially true, who denies that it is really or indeed even approximately correct, is surely obliged to provide an alternative explanation for the impressive success of science in matters of prediction and control. And this can be done.

Note, to begin with, that the dogma that successful prediction and application betoken the truth of science is belied by all those innumerable cases where clearly false beliefs induce people to do or to expect the right thing. It is simply not correct that our beliefs must actually be true – or at least "approximately true" or "roughly right" – to provide for effective prediction or praxis. For centuries sailors navigated successfully by the stars on the basis of thinking them to be "fixed". But this is certainly not even "approximately" true: neither is it the case that most stars are altogether fixed nor that all stars are nearly fixed. Bertrand Russell's criticism of the pragmatists provides various quaint object lessons, such as this:

> Dr. Dewey and I were once in the town of Changsha during an eclipse of the moon; following immemorial custom, blind men were beating gongs to frighten the heavenly dog, whose attempt to swallow the moon is the cause of the eclipses. Throughout thousands of years, this practice of beating gongs has never failed to be successful; every eclipse has come to an end after a sufficient prolongation of the din.[2]

Successful prediction and application can obviously attend false claims as well as true ones.

It is doubtless sensible to endorse the implication thesis: "If p is true, then we shall be at least as successful in implementing p as in implementing any of its false alternatives p', p'', etc." But, of course, its converse does not follow from this thesis. (To move to the conclusion that success underwrites truth is to commit the fallacy of affirming the consequent.)

Correctness is (most fortunately) not essential for success in the utilization of our beliefs. We can apply Euclidean geometry in terrestrial mensuration with great success. But does that mean that physical space is Euclidean rather than Reimannian or Lobatchewskian? Not at all! It simply means that in making our measurements we are generally operating below the threshold where the differences between these systems are sufficient to make themselves felt through frustrating effective mensuration.

In our interactions with nature, she puts questions to us which we may answer either correctly or incorrectly. But we are in the happy position that it can readily transpire that all goes well in circumstances where we are far off the mark – where success in applications prevails despite failures of "correctness" in truth telling. You need nourishment. And so nature confronts you with something and asks "Is it edible?" in circumstances where success hinges on getting *this* right. You say "yes" because you think it is a pear. You are wrong – it is an apple. But since apples too are edible, "all's well that ends well". It is certainly *not* the case that you are "approximately" right with regard to the item at issue – that a pear is *approximately* an apple. But the crucial point is that you are wrong in ways that are (*ex hypothesi*) immaterial to the application in question—this item is edible. Your error just does not matter for the question in view.

To be sure, we are often told that we must accept the commitments of a successful scientific theory lock, stock and barrel in order to account for its success. J. J. C. Smart, for example, reasons as follows. Suppose that we have a successful scientific theory T which assumes unobservable micro-objects and a phenomenalized version of T, T', which asserts are only those claims that follow from T strictly at the level of observable macro-phenomena. Then, so Smart maintains,

I would suggest that the realist could (say) . . . that the success of T' is explained by the fact that the original theory T is true of the things that it is ostensibly about; in other words by the fact that there really are electrons or whatever is postulated by theory T. If there were no such things, and if T were not true in a realist way, would not the success of T' be quite inexplicable? One would have to suppose that there were innumerable lucky accidents about the behavior mentioned in the observational vocabulary, so that they behaved miraculously *as if* they were brought about by the non-existent things ostensibly talked about in the theoretical vocabulary.[3]

On such a view, a scientific theory that underwrites success in prediction and control thereby assures the actual existence of the objects it envisions. Applicative success betokens ontological accuracy.

This plausible-seeming view runs into major difficulties. The history of science is one long litany of abandoned "truths" and abandoned "objects":

- The crystalline spheres of ancient and medieval astronomy.
- The humoral theory of medicine.
- The effluvial theory of static electricity.
- The "catastrophist" geology, with its commitment to a universal (Noachian) deluge.
- The phlogiston theory of chemistry.
- The caloric theory of heat.
- The vibratory theory of heat.
- The vital force theories of physiology.
- The electromagnetic aether.
- The luminiferous aether.
- The Aristotelian theory of circular inertia.
- The putative process of spontaneous generation.[4]

And the era of scientific changes of mind is far from over. Scientific theories at the frontier of research have a notoriously short half-life. Our current theories do not replace old error by secured truth but by *conjectures* regarding the truth that will – so we fully expect – eventually come to be seen as gravely defective. We can be fairly certain that the things they envision do not exist exactly as they envision them.

The realization of success in the predictive and applicative implementation of our scientific beliefs has little to do with their actual correctness. One must reject the totally false analogy that nature is a kind of combination lock and that to "unlock its secrets" we must "get the matter right".

2. TRUTH IS NOT THE BEST EXPLANATION OF SUCCESS IN PREDICTION AND EXPLANATION

But surely – we shall be told – its actual truth, its correctly describing the real world, is the best possible reason for the practical efficacy of a belief. While factors other than the truth of a thesis can, in principle, explain its successful implementability, still, its truth would seem to afford the *best* explanation of such success. As one recent writer puts it:

> The history of a mature science like physics is marked by an impressive increase in the predictive power of theories. . . . The best explanation of this phenomenon is the realist hypothesis that theories are capturing more and more theoretical truth about the world. [Otherwise] . . . this increase in predictive power is simply a total mystery.[5]

And so it would seem that the inference from "successfully applicable" to "(probably) true" affords an appropriate transition thanks to the inductive principle of "inference to the best explanation", and thus serves to validate the truth, or the approximate truth, of natural science.

But is it indeed the case that a scientific theory's actually being true does really provide the best available explanation for its efficacy in matters of prediction and control? By no means! Suppose *your picture* of a certain real-world situation is as noted in the diagram below.

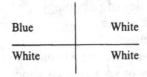

Fig. 1.

Is the best explanation of the fact that you can operate successfully with this description that the world is *really* made up in this way – that *everything* in the north-west quadrant is actually blue – that the world actually is as you believe it to be? Not at all! In actuality (so let us suppose) each quadrant is itself an aggregation of small rectangles, and closer inspection reveals that the indicated apparent color is actually only the *preponderant* color of those smaller components. As this suggests, the *best* explanation of the applicative efficacy of a belief is not its actual *truth* but simply that its departures from the truth do not make themselves felt so strongly at the prevailing level of operation as to invalidate our predictions and interactions *at this level*.

Its predictive and applicative success similarly does not show the *truth* of our theoretical picture of nature, but only its *local empirical indistinguishability* from the truth (whatever that may turn out to be) at the present level of observational sophistication. Our praxis within the parametric region in which we currently operate presumably works well in point of prediction and application simply because it operates at a level of crudity that avoids *detectable* errors. The most we can claim is that the inadequacies of our theories – whatever they may ultimately prove to be – are not such as to manifest themselves within the (inevitably limited) range of phenomena that currently lie within our observational horizons. This circumstance certainly has cognitive utility. But this is a far cry from justifying the claim that the theories that work well "as best we can tell" are thereby *correct*.

In general, the most plausible inference from the successful implementation of our factual beliefs is not that they are right, but merely that they (fortunately) do not go wrong in ways that preclude the realization of success within the applicative range of the particular contexts in which we are able to operate at the time at which we adopt them. Ptolemaic astronomy underwrote many successful applications, but this does not show that it is true. All it means is that there is a better theory (Newtonian astronomy) that explains how it was able to achieve those "unmerited" successes. Galenic medicine with its complex congeries of humors actually enjoyed many therapeutic successes in its prescriptions and treatments. But that does not establish its claims to truth. (All we can say is that there is a superior theory – modern scientific medicine – which can be used to explain how it is that those Galenic remedies were therapeutically useful.)

When we set out from the circumstance of a theory's applicative successes, then the "inference to the best explanation" leads not to this theory's truth, but to the thesis that this theory (T) is such that:

($\exists$ T') [(T' is more adequate than T) & (T' explains how T generates successful applications)]

While "it has actually got it right" is a *possible* explanation of why science "works" so effectively in prediction and control, the fact remains that this *possible* explanation is not the *best* explanation. An even better explanation proceeds from the far securer premiss that whatever failings and failures the science of the day might have simply do not manifest themselves at the current level of investigative sophistication. This explanation is better because its premiss, being less demanding, is more secure. In going for the *best* explanation of a family of successful implementations, the destination we reach is not "right" but merely "good enough for present purposes". Successful predictions and applications mean exactly no more than that – they betoken no more than pragmatic success under present working conditions and do not touch the issue of the actual descriptive correctness of those theories. The thesis that "The best explanation of the success of a scientific theory in matters of explanation, prediction, and control is that it is true" is simply false.

3. PRAGMATIC AMBIGUITY

Pierre Duhem made the sound and useful point that the falsification of a scientific prediction is a highly ambiguous item of information. For if a group of scientific theories and auxiliary hypotheses T_1 T_2, . . ., T_n collectively yield a prediction P that *fails* to be realized, then what we have is just the following pair of facts.

(1) $(T_1$ & T_2 & . . . & $T_n) \rightarrow P$
(2) $-P$

And from this we can conclude: $-(T_1$ & T_2 . . . & $T_n)$. All that we have is that something is wrong somewhere within the family: $T_1, T_2, \cdot \cdot \cdot ,$ T_n. But we have no idea what is amiss; we can make no particular imputation of fault. The lesson is straightforward. When things go wrong with a prediction to which various theories contribute, we cannot tell specifically where to attribute the blame.

But the reverse side of this reasoning is just as valid. A successful prediction is also a highly ambiguous datum. For if a group of scientific theories and auxilary hypotheses, $T_1, T_2, \ldots, T_n$, collectively yield a *successful* prediction P, then what we have is just this pair of facts.

(1) $(T_1 \& T_2 \& \ldots \& T_n) \rightarrow P$
(2) P

Short of committing the fallacy of affirming the consequent, we can conclude little or nothing about those T_is themselves, and strictly speaking, all that we can conclude is as follows.

$$(\exists Q) [(T_1 \& T_2 \& \ldots \& T_n) \rightarrow Q] \& Q \& [Q \rightarrow P])^6$$

Thus all we know when those two aforementioned facts are given is that there is some truth lurking somewhere amongst the consequences of T_1, $T_2, \ldots, T_n$ which is responsible for prediction Ps being true. The lesson is again straightforward. When things go right with a prediction to which various theories contribute, we cannot tell specifically where to attribute responsibility for this success. All that we know is that some truth or other entailed by the T_i –conjunction is in a position to assure P – that something or other both true and P–yielding is assured by the T_i –conjunction. When we meet with predictive success in applying one of our scientific theories, the conclusion we can draw is *not* that the theory itself gets the matter right, but that there is something other about it that is true – some consequence of it regarding whose nature we have no further information whatsoever.

The Duhem case of a failed prediction and this contrast case of a successful one run altogether parallel. When all we know is that a prediction underwritten by a group of theses fails, we have no idea of just where to lay the blame. Analogously, when all we know is that a prediction underwritten by a group of thesis succeeds, we have no idea just where to give the credit.

Its successful implementability does not indicate the (actual or probable or approximate) truth of a scientific theory. We do not have that the successful implementation of a belief p entails its truth or even its probable or approximate truth:

(1)   (p yields success) $\rightarrow$ 't'(p) $\lor$ probably ['t' (p)] $\lor$ approximately [$T(p)$]]. (Here 't' represents truth.)

Rather, we can get no further than the idea that there is some (conceiv-

ably wholly unknown and unthought-of) better theory which explains
this success.

(2)   (p yields success) $\rightarrow$ ($\exists\, q$) [$q > p$ & qE (p yields success)].
           (Here ">" represents probative superiority and "E" rep-
           resents "provides a good explanation for".)[7]

And the crucial fact is that this superior theory may itself *not* belong to
the range of our current scientific knowledge at all – and may even be
incompatible with it! It may lie wholly outside our ken in just the way in
which we say that Ptolemy's astronomy and Stahl's chemistry had
various successful applications which we can account for on a basis
altogether different from anything envisioned in these theories. The
successful application of a scientific theory does no more than indicate
that there is some superior alternative (conceivably one that is *very*
different from this theory itself) which accounts for its applicative
success.

All that we can ever conclude from the sort of successes that scientific
theories of the day put at our disposal is that whatever theory we may
eventually adopt as more adequate than the one currently in hand will
afford a deeper way of understanding that (and why), mistaken ideas to
the contrary notwithstanding, we proceed successfully within a limited
range of operation that is currently accessible. When a scientific theory
of ours underwrites successful applications, we can only conclude that
there is an adequate explanation of why this result is obtained. But that
this explanation lies in the fact that the theory is true and correct is
something we are certainly not entitled to maintain.

Thus, when a theory is "confirmed" through the success of its applications, we cannot appropriately conclude that it is (likely to be) true. We
do not know that the credit belongs to the theory at all – all we can
conclude is that an otherwise unspecifiable aspect of the overall situation can take the credit for the achievement of success – and emphatically not the whole theory as it stands. Columbus found land on the far
side of the Atlantic as he had expected to – but certainly not because his
theories were correct that China and the Indies lay some 3 000 to 4 000
miles across from Europe.

An applicatively successful theory need not present the truth, nor
need an applicatively better theory be more probably true. The relationship is far more oblique. The successful theory has some (otherwise unidentifiable) kernel of truth about it, and the superior theory
manages to enlarge this kernel. But this says nothing whatsoever about

the truth or probable truth or approximate truth of those theories themselves.[8]

4. THE LESSON

As these deliberations indicate, we have to uncouple their pragmatic efficacy from the issue of the truth and accuracy of our current theories. While we are certainly able to deploy our theories with impressive pragmatic efficacy at the present state-of-the-art of scientific operations (certainly far more effectively than before), we are not entitled to claim that current science has "got it right" – be it fully or even approximately.

Accordingly, the success of the applications of our current science does not betoken its unqualified truth or ultimate adequacy. All it indicates is that those various ways (whatever they may be) in which it doubtless fails to be true are not damaging to the achievement of these good results – that, in the context of those particular applications that are presently at issue, its errors lie beneath the penalty level of actual failure.

The fact is that nature can be error-tolerant to a substantial degree. As the situation in Figure 2 suggests, the applicative success of our scientific theories indicates no more than that we frequently find ourselves in circumstances where (mercifully) the penalty of error is relatively small. Thus, there is no good reason to think that even scientific theories that are impressively "well confirmed" through applicative successes are therefore *true*.

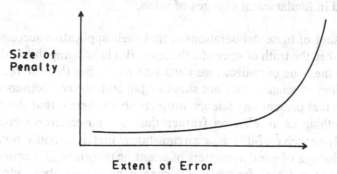

Note: In situations of this sort, substantial increases in error may involve but little loss in success – the errors may exact only very small penalties.

Fig. 2.   A common correlation between error in belief and penalty of implementation.

In the light of such considerations we can hardly take the stance of a naive scientific realism which holds that the world actually is as current science takes it to be, and that the theoretical entities envisioned by current science actually exist as current science envisions them to be. The sort of realism we can adopt is (at best) one of the estimationist type. There is no problem in saying that natural science *seeks* to get at "the real truth" of things – just in claiming that it *succeeds* in this aim. There is no problem in saying that natural science becomes increasingly *less inadequate* – just in saying that it becomes increasingly *more correct*. All we can properly infer from the applicative efficacy of our scientific theories is that the ways in which they go wrong do not prevent our achieving success; that they are *right* is something we are not entitled to claim. The successful application of a body of science does not show that all its theoretical statements, or most of them, or indeed any of them, are actually or even approximately true in their universally lawful guise.

The move from the claim that "our science enables us to predict and control with substantial success" to the claim that "our science delivers into our hands theories that adequately present some significant measure of actual truth about the world" is altogether unwarranted. All we can maintain on behalf of our present scientific theories is that over the range of situations currently within our purview they work better than any alternative available to us. But that they are *correct* is something else altogether. We can never rest satisfied on the basis of applicative success that "we have got it right" – be it fully or approximately – for we must remain alive to the realization that at deeper levels of scientific understanding we may (nay, probably will) in due course become involved in fundamental changes of mind.

The upshot of these deliberations is that their applicative success does *not* betoken the truth of scientific theories. But in bringing this finding to bear on the issue of realism, we must also remember that the reality of the entities of science does not stand or fall with the correctness of the theories that present-day science projects about them – that the reality of something *as an electron* (rather than as "an electron-just-as-we-currently-conceive-of-it") is a circumstance that is broadly invariant under changes of mind in matters of scientific theorizing. Of course, we cannot say *just how far* we could change our views about electrons without giving up our willingness to continue to characterize as "electrons" the entities we currently describe as such – any more than we can say just how far evolutionary change in sparrows could proceed before we would cease calling them "sparrows". The salient point is that a prudent doubt about the correctness of our scientific theories can be accepted without compromising the lessons for scientific realism drawn in the preceding chapter's deliberations regarding "schoolbook science".

On this basis, the contention that applicative efficacy does not betoken theories' truth is common ground between the present position and that of recent instrumentalists. But the anti-realistic conclusion that they want to draw from this contention is the exact opposite of our own position. Adhering to the view that the existence of the theory-entities of science stands or falls with the correctness of scientific theories, they see the success-does-not-imply-truth view of theories as undermining realism. The present position, by contrast, is that the very fact that our theories regarding the entities and processes standard in science are (presumably) defective is one of the surest signs of the objective reality of the items with which natural science deals.

The conclusion is that our scientific theories are attended by applicative success not because they are correct, but because nature is "error tolerant" to a substantial degree. And so the gap between success and truth – vast as it is – does not give comfort to the anti-realist. The circumstance that our cognitive grasp of reality is imperfect in no way shows that there is not a reality there to be grasped. The deficiency of our knowledge is rather an argument for the mind-independence of the mind-independently real than an indication of non-existence. The fact that our cognitive grasp of reality is imperfect provides support for realism.

CHAPTER SEVEN

THE ANTHROPOMORPHIC CHARACTER OF HUMAN SCIENCE

SYNOPSIS. (1) Our scientific picture of nature is the product of an interaction to which both parties – we investigators and nature herself – make a crucial *and inseparable* contribution. (2) The inquiring intelligences of an extraterrestrial civilization might also develop a science. (3) But it would not necessarily be anything like our science. For while it unquestionably deals with the same world, it would doubtless differ in mode of formulation, subject-matter orientation, and in conceptualization. (4) The one-world one-science argument is ultimately untenable. (5) Natural science as we have it is a human artifact that is bound to be limited in crucial respects by the very fact of it being *our* science. (6) The world-as-we-know-it is *our* world – the projection on the screen of mind of a world-picture devised in characteristically human terms of reference. It is not that natural science cannot underwrite valid claims to realism, but rather that the reality of our science is a characteristically *human* reality.

1. SCIENTIFIC RELATIVISM

There is no adequate reason of general principle to think of our own human scientific view of the world as cognitively absolute – as devoid of any relativization to the character of the two-sidedly collaborative interaction that obtains between the world and its investigators. We must recognize that process and product are reciprocally interwoven, and that our scientific picture of nature is the product of an *interaction* in which both parties, nature and ourselves, make a formative contribution. The resulting account is one in whose overall constitution the respective inputs of the two parties simply cannot be separated from one another – at any rate by us.

The queries, "What is the discoverable character of nature? What are the detectable *components* of physical reality? And what are the discernible *regularities* that govern them?" pose ill-formed questions unless we first resolve: "By whom?" For the issue is one that is inevitably relativized to the nature-interactive resources and instrumentalities at the disposal of investigators. Certainly, the regularities of nature are something perfectly real and independent of the wants and wishes of inquirers. But nevertheless, their reality – as a *relational* reality – is a matter of interaction between the world and its investigators.

Consider an analogy. A thin thread of string floats on a pool of water. Suppose someone asks: "What is the shape (configuration) of that thread – in itself and without reference to the water on which it is emplaced?" We are baffled. What answer could one possibly give? There is only one shape, the interactive configuration in whose make-up the role of the water is every bit as determinative as the thread itself. The configurative disposition of the thread inheres in its interaction with the water – and the relative contributions of the two parties cannot be separated. Analogously, the "shape of our knowledge" in natural science is something interactive which hinges every bit as much on the medium of import (the inquirers and *their* constitution) as on the constitution of the natural environment itself.

Factual (empirically based) theses as to how things work in the world are always correlative with the issue of monitoring, of interaction/ detection, of what can be "perceived" about nature from the vantage point of being emplaced therein and equipped with certain facilities for interaction with it. And this circumstance makes for a certain sort of scientific relativism. Science is a matter of explaining "how things work" in the world. But we can only explain what we can discern, and our facilities for discernment reflect our mode of emplacement within nature's scheme of things. This inevitably conditions – and accordingly relativizes – the kinds of scientific issues we can even address.

On such a view, inquiry yields results that are inherently relational. It is not that there is no such thing as a self-subsistingly non-relational reality, but rather that "reality *as we get hold of it*" is a complex composite towards whose constitution we ourselves make an ineliminable contribution.

Such a view maintains that science furnishes appropriate information about the world, but information that is appropriate from our point of view. Science does not afford us a picture of "reality in itself", but rather is a matter of "reality as it presents itself to us inquirers of a certain particular sort".

If scientific information is thus indelibly relational – if natural science as we have it is in an important sense *our* science and "describes reality" not in a categorical and absolute way but by providing investigation-relative results at different levels of investigator-nature interaction – then we can hardly take the stance that science affords an account of "the objectively real facts" about the constitution and the process of nature. If our "scientific knowledge" is indeed process-relative to the nature of our inquiries, then those versions of scientific realism that envision a total absence of investigation-dependency in our scientific knowledge of the world are untenable.

To say all this is not to espouse scepticism and deny the attainability of duly evidentiated information about the world. Rather, it is to adopt relativism, to accept that we investigators always crucially condition the sort of information that our science is in a position to give us about the world – with the result that when our situation in point of data-accessibility and information-processing changes, so does the scientific world-picture at which we arrive.

What is thrown into question from this perspective is not the *existence* of "the real world" that is self-subsistingly mind-independent, but the *status of our knowledge* of it. For it emerges that this knowledge is not objectively absolute knowledge of mind-independent reality as such, but inquiry-relative knowledge of empirical reality – reality as accessible "from the human point of view". We arrive at the position that our knowledge of the world is developed from the characteristically human perspective in the overall disposition of things in nature. Our knowledge of reality is man-relativized because the reality with which it deals is *our* reality – nature as our cognitive instrumentalities reveal it to us.[1] Let us examine more closely the implications of this perspective.

2. THE PROBLEM OF EXTRATERRESTRIAL SCIENCE

Is natural science something altogether investigation-independent? Is it a body of self-subsistent fact, wholly disconnected from the practices and procedures of its practicing investigators, a destination that all sufficiently clever inquiring intelligences are bound ultimately to reach in common? To elucidate this issue, let us consider the prospect that a civilization of extraterrestrial aliens, living on a planet in some far-off galaxy, might also develop natural science.

This simple assumption is actually one of great complexity. This complexity relates not only to the actual or possible facts of the matter, but also – and crucially – to fundamental and deep-cutting questions about the very idea that is at issue here.

To begin with, there is the matter of just what it means for there to be another science-possessing civilization. Note that this is a question that *we* are putting – a question posed in terms of the applicability of *our* term "science". It pivots on the issue of whether *we* would be prepared to consider certain of *their* activities – once we understood them – as engaging in scientific inquiry. And would *we* be prepared to recognize what those aliens are doing as forming beliefs (theories) about how things work in the world and to acknowledge that they are involved in testing these beliefs observationally or experimentally and applying them in practical (technological) contexts? We must, to begin with, be prepared to accept those alien creatures as (non-human) *persons*, duly equipped with intellect and will, and we must then enter upon a complex series of claims with respect to their cognitive activities.

A scientific civilization is not merely one that possesses intelligence and social organization, but one that puts this intelligence and organization to work in a very particular way. This opens up a rather subtle issue of priority with regard to process versus product. Is what counts for a civilization's "having a science" primarily a matter of the substantive *content* of their doctrines (their belief structures and theory complexes)? Or is it primarily a matter of the *aims and purposes* with which their doctrines are formed?

The matter of content turns on the similarity of their scientific beliefs to ours, which is clearly something we would be ill-advised to count on. After all, the speculations of the nature-theorists of pre-Socratic Greece, our ultimate ancestors in the scientific enterprise, bear little resemblance to our present-day sciences, nor does the content of contemporary physics bear all that much resemblance to that of Newton's day. We would do better to give prime emphasis to matters of process and purpose.

Accordingly, the matter of these aliens "having a science" should be seen as turning not on the extent to which their *findings* resemble ours but on the extent to which their *project* resembles ours. The crucial issue is whether we are engaged in the same sort of rational inquiry in terms of the sorts of issues being addressed and the ways in which we are going about addressing them. The basic issue is not one of the *substantive similarity* of their "science" to ours but one of the *functional equivalence* of their projects to the scientific enterprise as we know it. Only if they are pursuing such goals as description, explanation, prediction, and control of nature will they be doing *science*.

3. THE POTENTIAL DIVERSITY OF "SCIENCE"

The pivotal question is: To what extent would the *functional equivalent* of natural science built up by the inquiring intelligences of an astronomically remote civilization be bound to resemble our science? In reflecting on this question and its ramifications, one soon comes to realize that there is an enormous potential for diversity.

To begin with, the *machinery of formulation* used in expressing their science might be altogether different. And given that the mathematical mechanisms at their disposal could be very different from ours, it is clear that their description of nature in mathematical terms could also be very different. (Not necessarily more true or more false, but just different.) Secondly, the *orientation* of an alien civilization's science might be very different. All their efforts might conceivably be devoted to the social sciences, to developing highly sophisticated analogues of psychology and sociology, for example. In particular, if the intelligent aliens were a diffuse assembly of units comprising wholes in ways that allow of overlap,[2] then the role of social concepts might become so paramount for them that nature would be viewed throughout in fundamentally social categories. Accordingly, their natural science might deploy explanatory mechanisms very different from ours. Communicating by some sort of "telepathy", they might devise a complex theory of empathetic thought-waves transmitted through an ideaferous aether.

Again, the aliens might scan nature very differently. Electromagnetic phenomena might lie altogether outside the ken of variant life-forms; if their environment does not afford them lodestones and electrical storms, the occasion to develop electromagnetic theory might never arise. The course of scientific development tends to flow in the channel of practical interests. A society of porpoises might lack crystallography but develop a very sophisticated hydrodynamics; one comprised of mole-like creatures might never dream of developing optics or astronomy. One's language and thought processes are bound to be closely geared to the world as one experiences it. As is illustrated by the difficulties we encounter in bringing the language of everyday experience to bear on subatomic phenomena, our concepts are ill-attuned to facets of nature different in scale or structure from our own. We can hardly expect a "science" that reflects such parochial preoccupations to be a universal fixture. The interests of creatures shaped under the remorseless pressure of evolutionary adaptations to very different – and endlessly variable – environmental conditions might well be oriented in directions very different from anything that is familiar to us.

Laws are detectable regularities in nature. But detection will, of course, vary drastically with the mode of observation, that is, with the sort of resources that different creatures have at their disposal to do their detecting. Everything depends on how nature pushes back against our senses and their instrumental extensions. Even if we detect everything we can, we will not have got hold of everything available to others. And the converse is equally true. Since the laws we find are bound to reflect the sorts of data we can get hold of, the laws that we (or anybody else) can manage to formulate will depend crucially on our place within nature – on how one is connected into its wiring diagram, so to speak.

A comparison of different civilizations' "sciences" here on earth suggests that it is not an outlandish hypothesis to suppose that the very *topics* of alien science might differ dramatically from those of ours. Consider the following features of our own case – for example, the fact that we live on the surface of the earth (unlike whales), the fact that we have eyes (unlike worms) and thus can *see* the heavens, the fact that we are so situated that the seasonal positions of heavenly bodies are intricately connected with agriculture – all these facts are clearly connected with the development of astronomy. The circumstance that those distant creatures would experience nature in ways radically different to ourselves means that they can be expected to raise very different sorts of questions. Indeed, the mode of emplacement within nature of alien inquirers might be so different as to focus their attention on entirely different aspects or constituents of the cosmos. If the world is sufficiently complex and multi-faceted, they might concentrate upon aspects of their environment that means nothing to us, with the result that their natural science is oriented in directions very different from ours.[3]

Then too, the *conceptualization* of an alien science might be very different from ours. For we must consider the theoretical possibility that a remote civilization might operate with a drastically different system of concepts in its cognitive dealings with nature. Different cultures and different intellectual traditions, to say nothing of different sorts of creatures, are bound to describe and explain their experience – their world as they conceive it – in terms of concepts and categories of understanding substantially different from ours. They would diverge radically with respect to what the Germans call their *Denkmittel* – the conceptual instruments they employ in thought about the facts (or purported facts) of the world. They could, accordingly, be said to operate with different conceptual schemes, with different conceptual tools used to "make sense" of experience – to characterize, describe and explain the items that figure in the world as they view it. The taxonomic and explanatory mechanisms by which their cognitive business is transacted might differ so radically from ours that intellectual contact with them would be difficult or impossible.

In order to clarify this position, consider the analogy between *perception* and *conception*. It is clear that there is no "uniquely correct and appropriate" mode of perception. Different sorts of creatures have different sorts of senses, and use them to perceive the world differently. The situation with conceptions is in many ways similar to this. To be sure, there is also an important disanalogy. Conception lets us step outside the domain of perception and enables us, as human scientists, to describe how very different sorts of creatures can sense the world – that is how they go about monitoring their environment in physical interaction in ways different from ours. We can represent their *sensory* framework within our *conceptual* framework. But this we cannot do with conception itself. There is no supra-conceptual level at which *we* (at any rate) can compare and contrast how different sorts of creatures might conceive the world. We ourselves must depict *any* sort of concepts in terms of *our* concepts – otherwise we transgress the limits of intelligibility. But this disanalogy just strengthens the point at issue, namely, that there is nothing necessarily definitive about *our* way of conceptualizing the world and that we have no choice but to accept that different sorts of creatures could do it very differently.

Epistemologists have often said things to the effect that people whose experience of the world is substantially different from our own are bound to conceive of it in very different terms. Sociologists, anthropologists, and linguists talk in much the same ways, and philosophers of science have recently also come to say the same sorts of things. According to Thomas Kuhn, for example, scientists who work within different scientific traditions, and thus operate with different descriptive and explanatory "paradigms", actually "live in different worlds".[4]

It is (or should be) clear that there is no simple, unique, ideally adequate concept-framework for "describing the world". The botanist, horticulturist, landscape gardener, farmer, and painter will operate from diverse cognitive "points of view" to describe one selfsame vegetable garden. It is mere mythology to think that the "phenomena of nature" can lend themselves to only one correct style of descriptive and explanatory conceptualization. There is surely no "ideal scientific language" that has a privileged status for the characterization of reality. Different sorts of creatures are bound to make use of different conceptual schemes for the representation of their experience. To insist on the ultimate uniqueness of science is to succumb to "the myth of the God's eye view". Different cognitive perspectives are possible, not one of them more adequate or more correct than any other when considered independently of the aims and purposes of their users.

Supporting considerations for this position have been advanced from very different points of view. One example is a *Gedankenexperiment* suggested by Georg Simmel in the last century, which envisaged an entirely different sort of cognitive being: intelligent and actively inquiring creatures (animals, say, or beings from outer space) whose experiential modes are quite different from our own.[5] Their senses respond rather differently to physical influences: they are relatively insensitive, say, to heat and light, but substantially sensitized to various electromagnetic phenomena. Such intelligent creatures, Simmel held, could plausibly be supposed to operate within a largely different framework of empirical concepts and categories; the events and objects of the world of their experience might be very different from those of our own: their phenomenological predicates, for example, might have altogether variant descriptive domains. In a similar vein, William James wrote:

Were we lobsters, or bees, it might be that our organization would have led to our using quite different modes from these [actual ones] of apprehending our experiences. It *might* be too (we can not dogmatically deny this) that such categories, unimaginable by us to-day, would have proved on the whole as serviceable for handling our experiences mentally as those which we actually use.[6]

The science of a different civilization would inevitably be closely tied to the particular pattern of their interaction with nature as funneled through the particular course of their evolutionary adjustment to their specific environment. The "forms of sensibility" of radically different beings (to invoke Kant's useful idea) are likely to be radically different from ours. The direct chemical analysis of environmental materials might prove highly useful, and bioanalytic techniques akin to our senses of taste and smell could be very highly developed, providing them with environmentally oriented "experiences" of a very different sort.

The constitution of alien inquirers – physical, biological and social – thus emerges as crucial for science. It would be bound to condition the agenda of questions and the instrumentalities for their resolution – to fix what is seen as interesting, important, relevant, and significant. Because it determines what is seen as an appropriate question and what is judged as an admissible solution, the cognitive posture of the inquirers must be expected to play a crucial role in shaping the course of scientific inquiry itself.

As long as the fundamental categories of thought – the modes of spatiality and temporality, of structural description, functional connection, and explanatory rationalization – are not seen as necessary features of intelligence as such, but as evolved cognitive adaptations to particular contingently constituted modes of emplacement in and interaction with nature, there will be no reason to expect uniformity. Sociologists of knowledge tell us that even here on earth, our Western science is but one of many competing world-pictures. When one turns outward toward space at large, the prospects of diversity become literally endless. It is a highly problematic contention even that beings constituted as we are and located in an environment such as ours must inevitably describe and explain natural phenomena in our terms. And with differently constituted beings, the basis of differentiation is amplified enormously. Our minds are the information-processing mechanisms of an organism interacting with a particular environment via certain particular senses (natural endowments, hardware) and certain culturally evolved methods (cultural endowments, software). With different sorts of beings, these resources would differ profoundly – and so would the cognitive products that would flow from their employment.

To clarify this idea of a conceptually different science, it helps to cast the issue in temporal rather than spatial terms. The descriptive characterization of *alien* science is a project rather akin in its difficulty to that of describing our own *future* science. It is a key fact of life that progress in science is a process of *ideational* innovation that always places certain developments outside the intellectual horizons of earlier workers. The very concepts in terms of which we think become available only in the course of scientific discovery itself. Like the science of the remote future, the science of remote aliens must be presumed to be such that we really could not achieve intellectual access to it on the basis of our own position in the cognitive scheme of things. Just as the technology of a more advanced civilization would be bound to strike us as magic, so its science would be bound to strike us as incomprehensible gibberish – until we had learned it "from the ground up". They might (just barely) be able to *teach* it to us, but they could not *explain* it to us by transposing it into our terms.

The most characteristic and significant difference between one conceptual scheme and another arises when the one scheme is committed to something the other does not envisage at all – something that lies altogether outside the conceptual range of the other. A typical case is that of the stance of Cicero's thought-world with regard to questions of quantum electrodynamics. The Romans of classical antiquity did not hold *different* views on these issues; they held no view at all about them. This whole set of relevant considerations remained outside their conceptual repertoire. The diversified history of terrestrial science thus gives one some miniscule inkling of the vast range of possibilities along these lines.

The "science" of different civilizations may well, like Galenic and Pasteurian medicine, simply *change the subject* in key respects so as to no longer "talk about the same things", but deal with materials (e.g., humors and bacteria, respectively) of which the other takes no cognizance at all. The difference with regard to "conceptual scheme" between modern and Galenic medicine is not that the modern physician has a different theory of the operation of the four humors from his Galenic counterpart, but that modern medicine has *abandoned* the four humors, and not that the Galenic physician says different things about bacteria and viruses but that he says *nothing* about them.

The more we reflect on the matter, the more decidedly we are led to the realization that our particular human conception of the issues of science must be presumed to be something parochial because we are physically, perceptually, and cognitively limited and conditioned by our specific situation with nature. Given intelligent beings with a physical and cognitive nature profoundly different from ours, one simply cannot assert with confidence what the natural science of such creatures would be like with respect to matters of substance. The natural science of an extraterrestrial civilization could differ drastically from ours, not in mode of formulation alone, but also in subject-matter orientation and thus, even more critically, in mode of conceptualization.

4. THE ONE-WORLD, ONE-SCIENCE ARGUMENT

One writer on extraterrestrial intelligence raises the question: "What can we talk about with our remote friends?" and answers with the remark: "We have a lot in common. We have mathematics in common, and physics, and astronomy".[7] Another writer maintains that "we may fail to enjoy their music, understand their poetry, or approve their ideals; but we can talk about matters of practical and scientific concern".[8] With respect to his hypothetical Planetarians, the ingenious Christiaan Huygens wrote, three centuries ago:

Well, but allowing these Planetarians some sort of reason, must it needs be the same with ours? Why truly I think 'tis, and must be so; whether we consider it as applied to Justice and Morality, or exercised in the Principles and Foundations of Science. . . . For the aim and design of the Creator is every where the preservation and safety of his Creatures. Now when such a reason as we are masters of, is necessary for the preservation of Life, and promoting of Society (a thing that they be not without, as we shall show) would it not be strange that the Planetarians should have such a perverse sort of Reason given them, as would necessarily destroy and confound what it was design'd to maintain and defend? But allowing Morality and Passions with those Gentlemen to be somewhat different from ours, . . . yet still there would be no doubt, but that in the search after Truth, in judging of the consequences of things, in reasoning, particulary in that fort which belongs to Magnitude or Quantity about which their Geometry (if they have such a thing) is employ'd, there would be no doubt I say, but that their Reason here must be exactly the same, and go the same way to work with ours, and that what's true in one part will hold true over the whole Universe; so that all the difference must lie in the degree of Knowledge, which will be proportional to the Genius and Capacity of the inhabitants.[9]

It is tempting to reason: "Since there is only one nature, only one science of nature is possible." Yet, on closer scrutiny, this reasoning becomes highly problematic. Above all, it fails to deal with the fact that while there indeed is only one world, nevertheless very different *thought-worlds* can be at issue in the elaboration of a "science".

It is surely naive to think that because one single object is in question, its description must issue in one uniform result. This view ignores the crucial impact of the describer's intellectual orientation. Minds with different concerns and interests and with different experiential backgrounds can deal with the selfsame items in ways that yield wholly disjoint and disparate results because different features of the things are being addressed. The *things* are the same, but their significance is altogether different. Perhaps it seems more plausible to argue thus: "Common problems constrain common solutions. Intelligent alien civilizations have in common with us the problem of cognitive accommodation to a shared world. Natural science as we know it is *our* solution of this problem. Therefore, it is likely to be *theirs* as well." But this tempting argument flounders on its second premiss. Their problem is not common with ours, because their situation must be presumed to be substantially different, since they live in a significantly different environment and come equipped with significantly different resources. In fact, to presuppose a common problem is to beg the question.

Science is always the result of *inquiry* into nature, and this is inevitably a matter of a *transaction* or *interaction* in which nature is but one party and the inquirer another. We must expect alien beings to question nature in ways very different from our own. On the basis of an *interactionist* model, there is no reason to think that the sciences of different civilizations will exhibit anything more than the roughest sorts of family resemblance.

Our alien scientific colleagues also scan nature for regularities, perforce using (at any rate, to begin with) the sensors provided them by their evolutionary heritage. They note, record, and transmit those regularities that they find to be useful or interesting, and then develop their inquiries by theoretical triangulation from this basis. Now, this is clearly going to make for a course of development that closely gears their science to their particular situation – their biological endowment ("their sensors"), their cultural heritage ("what is interesting"), and their environmental niche ("what is pragmatically useful"). Where these key parameters differ, we must expect that the course of scientific development will differ as well.

Presumably, there is only one universe, and its laws and materials are, as best we can tell, the same everywhere. We share this common universe with all life-forms. However radically we differ in other respects (in particular, those relating to environment, to natural endowments, and to style or civilization), we have a common background of cosmic evolution and a common heritage of natural laws. And so, if intelligent aliens investigate nature at all, they will investigate the same nature we ourselves do. But the sameness of the object of contemplation does nothing to guarantee the sameness of ideas about it. It is all too familiar a fact that when only human observers are at issue, very different constructions are often placed upon "the same" occurrences. As is clearly shown by the rival interpretations of different psychological schools – to say nothing of the court testimony of rival "experts" – there need be little uniformity in the conceptions held about one selfsame object from different "perspectives of consideration". The fact that all intelligent beings inhabit the same world does not countervail against the no less momentous fact that we inhabit very different ecological niches within it, engendering very different sorts of *modus operandi*.

The universality and intersubjectivity of our science, its repeatability and investigator-independence, still leave matters at the level of *human* science. As Charles Sanders Peirce was wont to insist, the aim of scientific inquiry is to allay *our* doubts – to resolve the sorts of questions we ourselves deem worth posing. Different sorts of beings might well ask very different sorts of questions.

There is no categorical assurance that intelligent creatures will *think* alike in a common world, any more than that they will *act* alike – that is, there is no reason why *cognitive* adaptation should be any more uniform that *behavioral* adaptation. Thought, after all, is simply a kind of action; and as the action of a creature reflects its biological heritage, so too does its mode of thought.

No one who has observed how very differently the declarations of a single text (the Bible, say, or the dialogues of Plato) have been interpreted and understood over the centuries – even by people of a common cultural heritage – can be hopeful that the study of a common object by different civilizations must lead to a uniform result. Like other books, nature is to some extent a mirror: what looks out depends on who looks in. Yet, such textual analogies are over-simplified and misleading, because the scientific study of nature is not a matter of decoding a preexisting text. There just is not one fixed basic text – the changeless "book of nature writ large" – which different civilizations can decipher in different degrees.

The development of a "science" – a specific codification of the laws of nature – always requires as input some inquirer-supplied element of determination. The result of such an interaction depends crucially on the contribution from both sides – from nature and from the intelligences that interact with it. A kind of "chemistry" is at work in which nature provides only one input and the inquirers themselves provide another – one that can massively and dramatically affect the outcome in such a way that we cannot disentangle the respective contributions of nature and the inquirer. Things cannot of themselves dictate the significance that an active intelligence can attach to them. Human organisms are essentially similar, but there is not much similarity between the medicine of the ancient Hindus and that of the ancient Greeks. After all, throughout the earlier stages of man's intellectual history, different human civilizations developed their "natural sciences" in substantially different ways. The shift to an extraterrestrial setting is bound to amplify this diversity. The "science" of an alien civilization may be even more remote from ours than the "language" of our cousin, the dolphin, is remote from our language. Perhaps reluctantly, we must face up to the fact that on a cosmic scale the "hard" physical sciences have something of the same cultural relativity that we encounter with the materials of the "softer" social sciences on a terrestrial basis.

Peirce wrote that truth is "the predestined result to which sufficient enquiry would ultimately lead."[10] And again: "let any human being have enough information and exert enough thought upon any question, and the result will be that he will arrive at a certain definite conclusion, which is the same as that which any other mind will reach under sufficiently favorable circumstances."[11] For Peirce, if there is any prospect of truth at all – if there is any "fact of the matter" – then all inquirers are ultimately destined to achieve agreement about it. But even if one grants this (surely problematic) thesis of an ultimate uniformity of result in *human* inquiry, there is no good reason to project this uniformity across species. Our "scientific truths" are not necessarily those of others.

Natural science – broadly construed as inquiry into the ways of nature – is something that is, in principle, endlessly plastic. Its development will trace out an historical course closely geared to the specific capacities, interest, environment, and opportunities of the creatures that develop it. We are deeply mistaken if we think of it as a process that must follow a route generally parallel to ours and issue in a roughly comparable product. It would be grossly unimaginative to think that either the journey or the destination must be the same – or even substantially similar.

Factors such as capacities, requirements, interests, and course of development affect the shape and substance of the science and technology of any particular space-time region. Unless we narrow our intellectual horizons in a parochially anthropomorphic way, we must be prepared to recognize the great likelihood that the "science" and "technology" of a remote civilization would be something *very* different from science and technology as we know it. Our human sort of natural science may well be *sui generis*, adjusted to and coordinated with a being of our physical constitution, inserted into the orbit of the world's processes and history in our sort of way. It seems that in science, as in other areas of human endeavor, we are prisoners of the thought-world that our biological and social and intellectual heritage affords us.

These considerations point to a clear lesson. Different civilizations composed of different sorts of creatures must be expected to create diverse "sciences". Each inquiring civilization must be expected to produce its own, perhaps ever-changing, cognitive products – all more or less adequate in their own ways but with little if any actual overlap in conceptual content. Though inhabiting the same physical universe with us, and subject to the same sorts of fundamental regularities, they must be expected to create, as cognitive artifacts, different depictions of nature, reflecting their different modes of emplacement within it.

5. THE ANTHROPROMORPHIC CHARACTER OF HUMAN SCIENCE

Natural science as it actually exists is a human artifact that is bound to be limited in crucial respects by the very fact of its being *our* science. A tiny creature living its brief life-span within a maple leaf could never recognize that such leaves are deciduous. The processes of this world of ours (even unto its utter disappearance) could make no cognitive impact upon a being in whose body our entire universe is but a single atom. No doubt the laws of our world are (part of) the laws of its world as well, but this circumstance is wholly without practical effect. Where causal processes do not move across the boundaries between worlds – where the levels of relevantly operative law are so remote that nothing happening at the one level makes any substantial impact on the other – there can be little if any overlap in "science". Science is limited to the confines of discernibility: as Kant maintained, the limits of our experience set limits to our knowledge.

A deep question arises: Is the mission of intelligence in the cosmos uniform or diversified? Two fundamentally opposed philosophical views are possible with respect to cognitive evolution in its cosmic perspective. The one is a *monism* that sees the universal mission of intelligence in terms of a certain shared destination, the attainment of a common cosmic "position of reason as such". The other is a *pluralism* that sees each intelligent cosmic civilization as forging its own characteristic cognitive destiny, and takes it as the mission of intelligence as such to span a wide spectrum of alternatives and to realize a vastly diversified variety of possibilities, with each thought-form achieving its own peculiar destiny in separation from all the rest. The conflict between these doctrines must in the final analysis be settled not by armchair speculation from general principles but by rational triangulation from the empirical data. This said, it must be recognized that the whole tendency of these present deliberations is toward the pluralistic side. It seems altogether plausible to see cognition as an evolutionary product that is bound to attune its practitioners to the characteristic peculiarities of their particular niche in the world-order.

There is, no doubt, a certain charm to the idea of companionship. It would be comforting to reflect that however we are estranged from them in other ways, those alien minds share *science* with us at any rate and are our fellow travellers on a common journey of inquiry. Our yearning for companionship and contact runs deep. It might be pleasant to think of ourselves not only as colleagues but as junior collaborators whom other wise minds might be able to help along the way. Even as many in sixteenth-century Europe looked to those strange pure men of the Indies (East or West) to serve as moral exemplars for sinful European man, so we are tempted to look to alien inquirers who surpass us in scientific wisdom to assist us in overcoming our cognitive deficiencies. The idea is appealing, but it is also, alas, unrealistic. In the late 1600s, Christiaan Huygens wrote:

For 'tis a very ridiculous opinion that the common people have got among them, that it is impossible a rational Soul should dwell in any other shape than ours. . . . This can proceed from nothing but the Weakness, Ignorance, and Prejudice of Men; as well as the humane Figure being the handsomest and most excellent of all others, when indeed it's nothing but a being accustomed to that figure that makes me think so, and a conceit. . . . that no shape or colour can be so good as our own.[12]

People's tendency to place all rational minds into a physical structure akin to their own is paralleled by a tendency to emplace all rational knowledge into a cognitive structure akin to their own. (Roland Pucetti even thinks that the fundamental legal and social concepts of extra-terrestrial societies must be designed on our lines![13])

Life on other worlds might be very different from the life we know. It could be based on a multivalent element other than carbon and be geared to a medium other than water, perhaps even one that is solid or gaseous rather than liquid. In his splendid book entitled *The Immense Journey*, Loren Eiseley wrote:

Life, even cellular life, may exist out yonder in the dark. But high or low in nature, it will not wear the shape of man. That shape is the evolutionary product of a strange, long wandering through the attics of the forest roof, and so great are the chances of failure, that nothing precisely and identically human is likely ever to come that way again.[14]

What holds for the material configurations of the human shape would seem no less applicable to the cognitive configuration of human thought. It is plausible to think that alien creatures will solve the problems of *intellectual* adjustment to their environment in ways as radically different from ours as those by which they solve the problems of physical adjustment. The physics of an alien civilization need resemble ours no more than does their physical therapy. We must be every bit as leery of *cognitive* anthropomorphism as of *structural* anthropomorphism. (Fred Hoyle's science-fiction story entitled *The Black Cloud* is thought-provoking in this regard.[15] The cloud tells a scientist what it knows about the world. The result is schizophrenia and untimely death for the scientist: the cloud's information is too divergent and dissonant.)

With respect to biological evolution it seems perfectly sensible to reason as follows.

What can we say about the forms of life evolving on these other worlds?. . . . [I]t is clear that subsequent evolution by natural selection would lead to an immense variety of organisms; compared to them, all organisms on Earth, from molds to men, are very close relations.[16]

The same situation will surely obtain with respect to cognitive evolution. The "sciences" produced by different civilizations here on earth – the ancient Chinese, Indians, and Greeks, for example – unquestionably exhibit an infinitely greater similarity than obtains between our present-day science and anything devised by astronomically remote alien civilizations. The idea of a comparison in terms of "advance" or "backwardness" is highly implausible. The prospect that some astronomically remote civilization is "scientifically more advanced" than ourselves – that somebody else is doing "our sort of science" *better* than we are – in the first instance requires that they are doing our sort of science at all. And this deeply anthropomorphic supposition is extremely dubious. (It should be stressed, however, that this consideration that "our sort of natural science" may well be unique is not so much a celebration of our intelligence as a recognition of our peculiarity.)

Empiricist philosophers of science sometimes worry whether the set of actual observations might fail to do justice to "the facts". They fret, for example, lest the characterization of reality be underdetermined by the actual observations. And they seek to remove these worries by turning from the *actual* to the totality of *all possible* observations. But, if "possible" here means "possible *for us humans*" – with our particular sort of observational technology – then it seems clear that we have to deal with one particular *modus operandi* of one particular sort of creature.

The chance that an alien civilization might carry science – that is, *our* sort of science – further than us is remote, to put it mildly. Aliens might well surpass us in many ways, but to worry (or hope) that they might surpass us in *science as we practice it* is to orient one's concern in an unprofitable direction.[17]

6. RELATIVISTIC INTIMATIONS

The ultimate reason why we cannot expect alien intelligences to be at work doing our sort of science is that the possible sorts of "natural science" are almost endlessly diverse. Sciences, understood as such in the functional-equivalency terms laid down above, are bound to vary with the cognitive instruments available in the physical constitution and mental equipment of their developers and with the cognitive focus of interest of their cultural perspective and conceptual framework. We must think of our sort of science as merely one alternative among others: our whole cognitive project is simply the intellectual product characteristic of one particular sort of cognitive life-form. There is good reason to see natural science as species-relative.

It follows that even if we could somehow manage (*per impossibile*) to bring *our* sort of science to perfection, there is no reason to think that it would yield the definitive truth about nature as any other inquiring being would ultimately conceptualize it. There is yet another whole dimension of "limitation" to which our natural science is subject – inherent in the consideration that (as best we can judge the matter) it represents a characteristically *human* enterprise. The world *as we know it* is *our* world – the projection on the screen of mind of a world-picture devised in characteristically human terms of reference. The world is real enough, independently of our ideas about it, but the-world-as-we-view-it is a construction of ours – correlative to our (characteristically human) place in the cosmic scheme.

Immanuel Kant's insight holds: there is good reason to think that natural science as we know it is not something universally valid for all rational intelligences as such, but a (partially) man-made creation that is, in crucial respects, correlative with our specifically human intelligence. We have little alternative to supposing that our science is limited precisely by its being *our* science. The inevitability of empiricism – the fundamentality of experience for our knowledge of the world – means that our scientific knowledge is always relativized ultimately to the kinds of experiences we can have. Our science is bound to reflect our nature – to be conditioned and delimited by the sorts of creatures we are with respect to our mode of "sensory" involvement in the world's scheme of things. The "scientific truth" that we discover about the world is *our* truth, not so much in the sense that "we make it up," but rather in the sense that it reflects our human mode of emplacement within nature.

The theses and theories of our science are necessarily based on "the available data" and, accordingly, reflect the character of our interactions with nature. An interaction is a two-sided process to which each party makes an essential contribution – and where the character of these respective contributions cannot be altogether distinguished and clearly separated. And the potential plurality of modes of judgment means that there is no single definitive way of knowing the world. It entails a radical cognitive Copernicanism which recognizes that our position is no more central in the cognitive than in the spatial order of things.

One recent writer sees the scientific realist as committed to the doctrine that:

a scientific theory (containing reference to objects not directly observable) purports to reveal what the world is really like in itself, as opposed to how it presents itself to us, with our particular observational capacities, conceptual equipment, and location in space and time.[18]

But this contention is very questionable indeed. For two quite different sorts of observer-independence are available with respect to theses:

(1)   No reference to observers is involved in any way, manner, or shape.

(2)   There is reference to observers, but only in the hypothetical/

evidential mode: "If someone proceeded in this-and-this-way he would (be bound to) arrive at such-and-such findings."

Claims of the second sort – that investigators who go at it in *our way* are destined to arrive at our results – are nevertheless perfectly objective, even though observer-invoking. And they are altogether respectable from a "scientific" point of view. There is nothing inherently unscientific about information that reflects how the world presents itself to us.

The conditionalized realism we have supported here is clearly a relativized realism. All the same, it is a perfectly good realism. And this conditionalized mode of realism offers us a great advantage in resolving the tension between the relativistic character of human inquiry and the realistic claims of natural science.

We face the inconsistency spawned by the following plausible-seeming theses:

(1)   The real truth about nature is unique and absolute.
(2)   Knowledge of the real truth about nature is available to inquirers.
(3)   One's knowledge is always relative to one's place in the cognitive scheme of things.
(4)   Different knowers (people, eras, cultures, civilizations) have different places in the cognitive scheme of things.

Here theses (3) and (4) entail the existence of many different orders of knowledge, while (1) and (2) demand the uniqueness of real knowledge. But in the final analysis, there is no escaping the "facts of life" reflected in (3) and (4). Thus only two routes to inconsistency resolution are open. One can deny thesis (1) and become a relativist, adopting a theory of distinct orders of truth. Or one can deny thesis (2) and enroll among the sceptics. This former essentially relativistic position clearly affords the preferable option, and this represents the line taken in the present discussion.

Such an approach leads to a realism all right, but one that is relativistic because of its insistence that any science will reflect its deviser's particular "slant" on reality in line with certain investigator-characteristic modes of interaction with nature. On such a view, knowledge of reality is always (in some crucial respect) cast in terms of reference that reflect its possessor's cognitive proceedings. There is, certainly, a mind-independent reality, but cognitive access to it is always mind-conditioned. All that we can ever know of reality is mediated through conceptions that reflect *how this reality affects us*.

Thus, the resultant position is one of a relativism holding that "our knowledge" of the world always to some extent reflects the circumstance of its being *our* knowledge of the world. It is not that natural science can make good no valid claims to realism, but rather that the reality of our science is a characteristically *human* reality. (The version of scientific realism we adopt is thus a relativistic one which, as we shall consider in the closing chapter, also gives idealism its due.)

CHAPTER EIGHT

EVOLUTION'S ROLE IN THE SUCCESS OF SCIENCE

SYNOPSIS. (1) How is the impressively effective coordination of thought and reality that is provided by mathematicizing natural science to be explained? (2) Is it perhaps the case that this success is simply inexplicable? Surely not! (3) To account for the cognitive accessibility of nature, a two-sided explanation is needed – one in which both mind and nature must play a duly collaborative role. (4) Our own side of the story lies in the fact that mind is an evolved product of nature's operations. (5) Nature's side of the story lies in its providing the stage-setting for the evolutionary development of mathematicizing intelligence. (6) On this basis, we can obtain a scientific explanation of how science itself is possible. (7) The character of this explanation is not such, however, as to support the claim that "science has it right". The critical fact is that nature must be "error tolerant" in order to provide for cognitive evolution to occur at all. And this means that we can account for the extraordinary success of natural science short of maintaining its actual correctness.

1. THE PROBLEM OF MIND/REALITY COORDINATION

If one rejects (as one should) the idea that the natural science of the day delivers into our hands the real, or indeed even the "approximate" truth of things, then how can one possibly account for the enormous applicative success of science in matters of prediction and control? Is it not a grave impediment to scientific fallibilism that in abandoning the notion that science "gets it right" one loses all explanatory grip on its extraordinary success in achieving those impressively wide-ranging and detailed explanations, predictions, and interventions in nature? Here, clearly, is a significant challenge that must be met by any fallibilistic rejection of science-gets-it-right realism.

For factual knowledge to arise at all, the beliefs of inquiring minds and the world's actual arrangements must be duly coordinated in mutual attunement. But when two parties agree, this can come about in very different ways. Consider two piles of rocks that correspond in point of size. Clearly, this might transpire because:

(i)   (1) is adjusted to (2). (The size of (2) is the independent variable and that of (1) the dependent variable.)

(ii)   (2) is adjusted to (1).

(iii)   There is a two-way coordination, a reciprocal adjustment, a give-and-take coupling, an *interaction*.

Exactly the same spectrum of possibilities exists with respect to the issue of mind/reality coordination. Here too, there are three alternatives:

(1)   *Physicalism or ontological materialism*: In human knowledge, mind and extra-mental reality agree because extra-mental reality constrains the operations of mind along a "one-way street" with mental processes as the causal products of an extra-mental reality.

(2)   *Ontological idealism*: In human knowledge, mind and extra-mental reality agree because mind actually *constructs* that seemingly "extra-mental" nature. Mind is in control, so that, . in consequence, all else agrees with mind.

(3)   *Interactionism*: In human knowledge there is agreement between mental operations and extra-mental reality through a mutual accommodation engendering a process of give-and-take interaction, in the course of which our conceptions are coordinated with the ways of extra-mental reality through the operation of evolutionary processes.

The third alternative clearly provides the most attractive option here – and the most viable one. To see why this is so let us examine more closely how the mind/nature coordination essential to a knowledge of natural fact actually arises.

2. THE COGNITIVE ACCESSIBILITY OF NATURE

How is natural science – and, in particular, physics – possible at all? How is it that we humans, mere dust-specks on the world's immense stage, can manage to unlock nature's secrets and gain access to her laws? And how is it that our mathematics – seemingly a free creative invention of the human imagination – can be used to characterize the *modus operandi* of nature with such uncanny efficacy and accuracy? Why is it that the majestic lawful order of nature is intelligible to us humans in our man-devised conceptual terms?[1]

As long as people thought of the world as the product of the creative activity of a mathematicizing intelligence, as the work of a creator who proceeds *more mathematico* in the design of nature, the issue is wholly unproblematic. God endows nature with a mathematical order and mind with a duly consonant mathematicizing intelligence. There is thus no problem about how the two get together – God simply fixes it that way. But, of course, if *this* is to be the canonical rationale for the human mind's grasp on nature's laws, then in foregoing explanatory recourse to God we also – to all appearances – lose our hold on the intelligibility of nature.

Accordingly some of the deepest intellects of the day think that our hold on nature's intelligence is gone forever. Various scientists and philosophers of the very first rank nowadays confidently affirm that we cannot hope to solve this puzzle of the intelligibility of nature in a mathematically lawful manner. Erwin Schroedinger characterizes the circumstance that man can discover the laws of nature as "a miracle that may well be beyond human understanding".[2] Eugene Wigner asserts that "the enormous usefulness of mathematics in the natural sciences is something bordering on the mysterious, and there is no rational explanation for it"[3]; he goes on to wax surprisingly lyrical in maintaining that "The miracle of the appropriateness of the language of mathematics for the formulation of the laws of physics is a wonderful gift which we neither understand nor deserve."[4] Even Albert Einstein stood in awe before this problem. In a letter written in 1952 to an old friend of his Berne days, Maurice Solovine, he wrote:

You find it curious that I regard the intelligibility of the world (in the measure that we are authorized to speak of such an intelligibility) as a miracle or an eternal mystery. Well, *a priori* one should expect that the world be rendered lawful only to the extent that we intervene with our ordering intelligence . . . [But] the kind of order, on the contrary, created, for example by Newton's theory of gravitation, is of an altogether different character. Even if the axioms of the theory are set by men, the success of such an endeavor presupposes in the objective world a high degree of order that we were *a priori* in no way authorized to expect. This is the "miracle" that is strengthened more and more with the development of our knowledge. . . . The curious thing is that we have to content ourselves with recognizing the "miracle" without having a legitimate way of going beyond it . . .[5]

According to all these eminent physicists we are confronted with a genuine mystery. We have to acknowledge *that* nature is intelligible, but we have no prospect of understanding *why* this is so. The problem of nature's intelligibility through man's mathematical theorizing is seen as intractable, unresolvable, hopeless. All three of these distinguished Nobel laureates in physics unblushingly employ the word "miracle" in this connection.

Perhaps, then, the question is even illegitimate and should not be raised at all. Perhaps the issue of nature's intelligibility is not just intractable, but actually *inappropriate* and somehow based on a false presupposition. For to ask for an explanation as to *why* scientific inquiry is successful presupposes that there *is* a rationale for this fact. And if this circumstance is something fortuitous and accidental, then no such rationale exists at all. Just this position is preferred by various philosophers. For example, it is the line taken by Karl Popper, who writes:

[Traditional treatments of induction] all assume not only that our quest for [scientific] knowledge has been successful, but also that we should be able to explain why it is successful. However, even on the assumption (which I share) that our quest for knowledge has been very successful so far, and that we now know something of our universe, this success becomes [i.e. remains] miraculously improbable, and therefore inexplicable; for an approach at an endless series of improbable accidents is not an explanation. (The best we can do, I suppose, is to investigate the almost incredible evolutionary history of these accidents . . .).[6]

Mary Hesse, too, thinks that it is inappropriate to ask for an explanation of the success of science "because science might, after all, be a miracle".[7] On this sort of view, the question of the intelligibility of nature becomes an illegitimate pseudo-problem – a forbidden fruit at which sensible minds should not presume to nibble. We must simply remain content with the fact itself, realizing that any attempt to explain it is foredoomed to failure because of the inappropriateness of the very project.

And so, on this grand question of how the success of natural science is possible at all, some of the shrewdest intellects of the day avow themselves baffled, and unhesitatingly proceed to shroud the issue in mystery, or incomprehension. Surely, however, such a view is of questionable merit. Eminent authorities to the contrary notwithstanding, the question of nature's intelligibility through natural science is not only interesting and important, but is also one which we should, in principle, hope to answer in a more or less sensible way. Surely, this issue needs and deserves a strong dose of demystification.

3. A CLOSER LOOK AT THE PROBLEM

How is it that we can make effective use of mathematical machinery to characterize the *modus operandi* of nature? How is mathematical exactness possible save under the supposition that "science gets it right"?

The pure logician seems to have a ready answer. He says: "Mathematics *must* apply to reality. Mathematical propositions are strictly *conceptual* truths. Accordingly, they hold of *this* world because they hold of *every possible* world." But this response misses the point of present concerns. Admittedly, the truths of *pure* mathematics obtain in and of every possible world. But they do so only because they are necessary truths and descriptively empty – wholly non-committal on the substantive issues of the world's operations. Their very conceptual status means that their propositions are beside the point of our present purposes. It is not the *a priori* truth of pure mathematics that concerns us, its ability to afford truths of reason. Rather, what is at issue is the *empirical applicability* of mathematics – its pivotal role in framing the *a posteriori*, contingent truths of lawful fact – which renders nature's ways amenable to inquiring reason.

After all, it is perfectly clear that the availability of pure mathematics in a world does not mean that this world's *laws* have to be characterizable in relatively straightforward mathematical terms. It does not mean that nature's operations have to be congenial to mathematics and graspable in terms of simple, neat, elegant, and rationally accessible formulas. In short, it does not mean that the world must be mathematically tractable – must be "mathematophile" in being receptive to the sort of descriptive treatment it receives in mathematical physics.

How, then, are we to account for the fact that the world appears to us to be so eminently intelligible in the mathematical terms of our natural science?

The answer to this question of the cognitive accessibility of nature to mathematicizing intelligence has to lie in a somewhat complex, two-sided story in which both sides, intelligence and nature, must be expected to have a part. Let us trace out this line of thought.

4. "OUR" SIDE

Our human side of this bilateral story is relatively straightforward. After all, *Homo sapiens* is an integral part of nature. We are connected into nature's scheme of things as an intrinsic component thereof – courtesy of the processes of evolution. And the kind of mathematics we are going to devise is pretty much bound to be the kind that is applicable. The experience we secure is inevitably an experience *of nature*. (That after all is what "experience" is – our intelligence-mediated reaction to the world's stimulating impacts upon us.) So the kind of mathematics – the kind of theory of periodicity and structure – that we devise in the light of this experience is the kind that is, in principle, applicable to nature as we can experience it.

Our mathematics is destined to be congenial to nature because it itself is a natural product; it fits nature because it reflects the way we are emplaced within nature as integral constituents thereof. Our intellectual mechanisms – mathematics included – fit nature because they are a product of nature's operations as mediated through the cognitive processes of an intelligent creature, which uses its intelligence to guide its interaction with a nature into which it is fitted in a certain sort of way.

As we considered in the preceding chapter, the mathematics of an astronomically remote civilization whose experiential resources differ from ours might well be substantially different from mathematics as we ourselves know it. Their dealings with quantity might be entirely anumerical – purely comparative, for example, rather than quantitative. Especially if their environment is not amply endowed with solid objects or stable structures congenial to measurement – if, for example, they were jellyfish-like creatures swimming about in a soupy sea – their "geometry" could be something rather strange, largely topological, say, and geared to flexible structures rather than fixed sizes or shapes. Digital thinking might go undeveloped, while certain sorts of analogue reasoning might be highly refined. In particular, if the aliens were assemblages of units social concepts might become paramount in their thinking, with those aggregates we think of as *physical* structures contemplated by them in *social* terms. Communicating by some sort of variable odors or otherwise "exotic" signals, they might, for example, devise a theory of thought-wave transmittal through an aether. Overall, the processes that underlie their mathematicizing might be very different indeed.

Mathematics is the theory of imaginable structures; and "imaginable" here is a matter of imaginable by a nature-evolved and nature-embraced creature. Admittedly, mathematics is not a natural science but a theory of noncontingent possibilities. Nevertheless, these possibilities are the ones conceived *by us* – by a being who conceives his possibility with a nature-evolved and nature-implanted mind. It is thus not surprising that the sort of mathematics we contrive is the sort of mathematics we find applicable to the conceptualization of nature. After all, the intellectual mechanisms we devise in coming to grips with the world – in transmuting sensory interaction with nature into intelligible experience – have themselves the aspect (among many other aspects) of being nature's contrivances in adjusting to its ways a creature it holds at its mercy.

One possible misunderstanding must be averted here. It is not being claimed that the development of mathematics is survival-conducive as such, that mathematics is a practical resource akin to food or shelter. The bearing of evolution is far more indirect than that. Mathematics is the natural product and expression of certain capabilities (of synthetic representation) and certain interests (the impetus to understanding) which themselves are survival-conducive. It is not mathematics that is of evolutionary instrumentality but the cognitive resources and interest that provide the building blocks by whose means we erect its structure.

It is no more a miracle that the human mind can understand the world than that the human eye can see it. The critical step is to recognize that the question: "Why do our conceptual methods and mechanisms fit 'the real world' with which we interact intellectually?" simply does not permit any strictly aprioristic answer in terms of purely theoretical grounds of general principle. Rather, it is to be answered in basically the same way as the question: "Why do our bodily processes and mechanisms fit the world with which we interact physically?" Both are alike and can be resolved in essentially evolutionary terms. It is no more surprising that man's mind grasps nature's ways than that man's eye can accommodate nature's rays or his stomach nature's food. Evolutionary pressure can take credit for the lot – though both biological and cultural evolution come into it. There is nothing "miraculous" or "lucky" in our possession of efficient cognitive faculties and processes. If we did not possess these, we just would not be here as inquiring creatures emplaced in nature thanks to evolutionary processes.

All the same, it could be the case that we do well as regards our cognitive adjustment only in the immediate local microenvironment that defines our particular limited ecological niche. The possibility still remains open that we do not really do all that well in a wider sense – that we get hold of only a small and peripheral part of a large and impenetrable whole. And so, man's own one-sided contribution to the matter of nature's intelligibility cannot be the *whole* story regarding the success of science. Nature too must do its part.

To clarify this issue we must move on to contemplate nature's contribution to the bilateral mind/nature relationship.

5. NATURE'S SIDE

What needs to be shown for present purposes is not just that mathematics is of *some* utility in understanding the world, but that it is bound to be of *very substantial* utility. We need to assure ourselves that the employment of mathematics can provide intelligent inquirers with an adequate and accurate grasp of nature's ways. Thus, we must dig more deeply into the issue of nature's amenability to inquiry and its "cooperation" with the probes of intelligence.

Certainly, the effective applicability of mathematics to the description of nature is there to some extent because we actually devise our mathematics to fit nature through the mediation of experience. However, the fact remains that man is an inquiring being emplaced within nature and forms mathematicized conceptions and beliefs about it on the basis of physical interaction with it in order to achieve a reasonably appropriate grasp of its workings. Thus nature too must "do its part": it must be duly benign. Obviously it must permit the evolution of inquiring beings. And to do this, it must present them with an environment that affords sufficiently stable patterns to make coherent "experience" possible, enabling them to derive appropriate *information* from the structured interactions that prevail in nature at large.

Nature's own contribution to the issue of the mathematical intelligibility of nature must be the possession of a relatively simple and uniform law structure – one that encodes so uncomplicated a set of regularities that even a community of inquirers possessed of only moderate capabilities can be expected to achieve a fairly good grasp of it.

But how can one establish that nature simply "*must*" have a fairly straightforward law structure? Are there any fundamental reasons why the world that we investigate by the use of our mathematically informed intelligence should operate on relatively simple principles readily amenable to mathematical characterization? There are indeed. For a world in which intelligence emerges by standard evolutionary processess has to be pervaded by regularities and periodicities in organism-nature interaction. But this means that nature must be cooperative in a certain and very particular way – it must be stable enough and regular enough and structured enough for there to be "appropriate responses" to natural events that can be "learned" by creatures. If such "appropriate responses" are to develop, nature must provide "suitable stimuli" in a duly structured way. A knowable world must incorporate experientiable structures. There must be regular patterns of occurrence in nature that even simple, single-celled creatures can embody in their make-up and reflect in their *modus operandi*. Even the humblest organisms that possess only the most rudimentary proto-intelligence (say, snails and algae) must so operate that certain *types of stimuli* (patterns of recurrently discernible impacts) call forth appropriately corresponding *types of response* – that such organisms can "detect" a structured pattern in their natural environment and react to it in a way that is advantageous to them in evolutionary terms. Even the simplest creatures must swim in a sea of detectable regularities readily accessible to intelligence. Accordingly, a world in which any form of intelligence evolves will have to be a world that is congenial to mathematics.

Moreover, intelligence must give an "evolutionary edge" to its possessors. The world must encapsulate straightforward learnable patterns and periodicites of occurrence in its operations – relatively simple laws. The existence of such learnable "structures" of stability in natural occurrence means that there must be *some* useful role for mathematics, which, after all, is the abstract and systematic theory of structure-in-general. And so we may conclude that a world in which intelligence can develop by evolutionary processes *must* be a world amenable to understanding in mathematical terms.[8] Such a world must be one in which such beings will find considerable ammunition in endeavoring to "understand" the world. Galileo already came close when he wrote in his *Dialogues* that: "Nature initially arranged things her *own* way and subsequently so constructed the human intellect as to be able to understand her."[9]

The development of *life* and thereupon of *intelligence* in the world may or may not be inevitable; the emergence of intelligent creatures on the world's stage may or may not be surprising in itself. But once they are there, and once we realize that they got there thanks to evolutionary processes, it can no longer be seen as surprising that their efforts at characterizing the world in mathematical terms should be substantially successful. *A world in which creatures possessed of high intelligence emerge through the operation of evolutionary processes must be a highly intelligible world*.

The apparent success of human mathematics in characterizing nature is thus nowise amazing. It may or may not call for wonder that intelligent creatures should evolve at all. But that once they have safely arrived on the scene through evolutionary means, that they should be able to achieve success in the project of understanding nature in mathematical terms is only to be expected. An intelligence-containing world whose intelligent creatures came by this attribute through evolutionary means must be substantially intelligible in mathematical terms. On this line of deliberation, then, nature admits to mathematical depiction not just because it has laws (is a *cosmos*), but because it has *relatively simple laws*, and those relatively simple laws must be there because if they were not, then nature just could not afford a potential environment for intelligent life.

The strictly hypothetical character of this general line of reasoning must be recognized. It does not maintain that the world has to be simple enough for its modes of operation to admit of elegant mathematical representation by virtue of some sort of transcendental necessity. Rather, what it maintains is the purely conditional thesis that *if* intelligent creatures are going to emerge in the world by evolutionary processes, *then* the world must be mathematophile, with various of its processes amenable to mathematical representation.

It must be emphasized that this conditional fact is quite enough for present purposes. For the question faced is why we men should be able to understand the operations of the world's stage in terms of our mathematics. The conditional story at issue fully suffices to accomplish this particular job.

6. SYNTHESIS

Let us review the course of reasoning. The overall question of the intelligibility of nature has two sides: (1) Why is mind so well attuned to nature? (2) Why is nature so well attuned to mind? The preceding discussion has suggested that the answers to these questions are not all that complicated, at least at the level of schematic essentials. The crux is that the mind must be attuned to nature because it is an evolved natural product of nature's operations. And nature must be accessible to mind because mind managed to evolve by an evolutionary route.

For nature to be intelligible, then, there must be an alignment that requires cooperation on both sides. Perhaps the analogy of cryptanalysis will help. If *A* is to break *B*'s code, there must be due reciprocal alignment. If *A*'s methods are too crude, too hit and miss, he can get nowhere. But even if *A* is quite intelligent and resourceful, his efforts cannot succeed if *B*'s procedures are simply beyond his powers. (The cryptanalysts of the seventeenth century, clever though they were, could not successfully apply their investigative instrumentalities to a high-level naval code of World War II vintage.) Analogously, if mind and nature were too far out of alignment – if mind were too "unintelligent" for the complexities of nature or nature too complex for the capacities of mind – the two just could not get into step. It would be akin to trying to rewrite Shakespeare with a 500 word vocabulary. We would lose too much important content. The situation would be like trying to keep tabs on a system containing ten importantly relevant degrees of freedom with a cognitive mechanism capable of keeping track of only four of them. If this were the case, mind could not accomplish its evolutionary mission of increasing survival probability. It would be better to adopt an alignment process that does not take the cognitive route.

The possibility of a mathematical science of nature is to be explained by the fact that, in the light of evolution, intelligence and intelligibility must stand in mutual coordination. Accordingly, the following three points are paramount:

(1)   Intelligence evolves within a nature that provides for life because it affords living creatures a good way of coming to terms with the world.

(2)   Once intelligent creatures evolve, their cognitive efforts are likely to have some degree of adequacy because evolutionary pressures align them with nature's ways.

(3)   It should not be surprising that this alignment eventually produces a substantially effective mathematical physics, because the fundamental structure of the operations of an intelligence-providing nature is bound to be relatively simple.

There may indeed be mysteries in this general area. (Questions such as: "Why should it be that *life* evolves in the world?" And, even more fundamentally: "Why should it be that the world exists at all?" Such questions have certainly been proposed as plausible candidates.) But be that as it may, the presently deliberated issue of why nature is intelligible to man, and why this intelligibility should incorporate a mathematically articulable physics, does not qualify as all that mysterious, let alone miraculous. No doubt, this general account is highly schematic and requires a great deal of amplification. A long and complex tale must be told about physical and cognitive evolution to fill in the details needed to put such an account into proper order. But there is surely good reason to hope and expect that a tale of this sort can ultimately be told. And this is the pivotal point. Even if one has doubts about the particular outlines of the evolutionary story we have sketched, the fact remains that *some such story* can provide a perfectly workable answer to the question of why nature's ways are intelligible to man in terms of mathematical instrumentalities. The mere fact that such an account is, in principle, possible shows that the issue need not be painted in the black or black of impenetrable mystery.

There is simply no need to join Einstein, Schroedinger, *et al.* in considering the intelligibility of nature as a miracle or as a mystery that passes all human understanding. If we are willing to learn from science how nature operates and how man goes about conducting his inquiries into its workings, then we should be able, increasingly, to remove the shadow of mystery from the problem of how a being of *this* sort, probing an environment of *that* type, and doing so by means of those evolution-arily developed cognitive and physical instrumentalities, should be able to arrive at a relatively workable account of how things work in the world. We should eventually be able to see it as only plausible and to be expected that inquiring beings should emerge and get themselves into a position to make a relatively good job of it. We can thus look *to science itself* for the materials that enable us to understand how natural science is possible. And there is no good reason to expect that it will let us down in this regard.[10]

7. IMPLICATIONS

Does such a scientific explanation of the success of science not explain too much? Does its account of the pervasiveness of mathematical exactness in science not indicate that "science gets it right" – a result that would contradict our historical experience of science's fallibilism? By no means. It is fortunate (and evolutionarily most relevant) that we are so positioned within nature that many "wrong" paths lead to the "right" destination – that flawed means can lead us to satisfying ends. If nature were a combination lock where we simply "had to get it right", and *exactly* right, to achieve success in implementing our beliefs, then we just would not be here. Evolution is not an argument that speaks unequivocally for the success of our cognitive efforts. On the contrary, properly construed, it is an indicator of our capacity to err and "get away with it".

The success of science can thus be understood on analogy with the success of the thirsty man who drank white grape juice, mistaking it for lemonade. It is not that he was roughly right – that grape juice is "approximately" lemonade – it is just that his beliefs were not wrong in ways that led to his being baffled in his purposes. Such defects as they have do not matter for the issues currently in hand. The "success" at issue one that is not unqualified but mixed.

Thus, we arrive at the picture of nature as an error-tolerant system. For consider the hypothetical situation of a belief-guided creature living in an environment that exacts a great penalty for "getting it wrong". Such a creature would have been eliminated long ago. It could not even manage to survive and reproduce long enough to learn about its environment by trial and error. If the world is to be a home for intelligent beings who develop in it through evolution, then it has to be benign. If *seeming* success in intellectually governed operations could not attend even substantially erroneous beliefs, then we cognizing beings who have to learn by experience just could not have made our way along the corridors of time. If nature were not error-forgiving, a process of evolutionary trial and error could not work and intelligent organisms could not have emerged at all. It follows that intelligence and the "science" it devises must pay off in terms of applicative success – irrespective of whether it manages to get things substantially right or not. In the circumstances, we cannot help laboring under the impression that our science is relatively successful.

Admittedly, applicative success calls for *some* alignment of thought-governed action with "the real *nature* of things", but only *some* – just enough to get by without incurring overly serious penalties in failure. Success here betokens not the presence of rightness but the absence of *egregious* error. The applicative success of our science is not to be explained on the basis of it actually getting at the real truth, but in terms of it being the work of a cognitive being who operates within an error-tolerant environment – a world-setting where applicative success will attend even theories that are substantially "off the mark". The applicative efficacy of science undoubtedly requires *some* degree of alignment between our world-picture and the world's actual arrangements – but just enough to yield the particular successes at issue.

The upshot of these deliberations, then, is that the success of science can be explained well short of the supposition that it manages to get at the real truth of things. It merely means that those ways (whatever they may be) in which it fails to be true are immaterial to the achievement of good results – that, in the context of the particular applications at issue, its inadequacies lie beneath the penalty level of actual failure. This matter of adequacy sufficient "to get by" – this idea of an "error-tolerant" nature permitting applicative success to attend factual falsity – is something that is in large measure explicable in evolutionary terms. For this sort of success can be (and doubtless is) the product of an evolutionary alignment of inquirers to the environment that they investigate – an alignment that can stop well short of a perfect fit simply because that world must, in the evolutionary nature of things, constitute a highly error-tolerant environment. This critical fact that nature is error-tolerant means that we can explain the extraordinary success of natural science without becoming involved in untenable claims about its correctness.

What we have here is a mitigated realism that is part and parcel of the systemic story told by natural science itself. It provides a coherent account which enables us to understand how it is that a creature constituted to all appearances as we are, and emplaced in a world constituted as this one (to all appearances) is, can use its evolutionarily given resources for cognitively probing its environment with substantial (apparent) efficacy.

Such an account is no doubt circular. It explains the possibility of our knowledge of nature on the basis of what we know of nature's ways. Its programmatic aim is to use the deliverances of natural science retro-spectively to provide an account of how an effective natural science is possible. Such a process is *not*, however, a matter of vitiating circularity, but one of a healthy self-sufficiency of our scientific knowledge that is, in fact, an essential part of its claims to adequacy.

CHAPTER NINE

THE ROOTS OF OBJECTIVITY

SYNOPSIS. (1) The facts about the things of this world are cognitively inexhaustible. (2) Real things – actually existing physical objects – have a cognitive depth whose bottom we cannot possibly plumb. (3) To have a correct conception of something we must have *all of its important properties* right. And this is something we generally cannot ascertain, if only because we cannot say what is really important before the end of the proverbial day. Accordingly, our conceptions of real things are inevitably corrigible. (4) This circumstance has a strongly realistic tendency. It means that the limits of *our* world – the world of our belief and of our information – cannot be claimed to be the limits of *the* world. It is the very limitation of our knowledge of things – our recognition that reality extends beyond the horizons of what we can possibly know or even conjecture about – that betokens the mind-independence of the real.

1. THE COGNITIVE INEXHAUSTIBILITY OF THINGS

From finitely many axioms, reason can generate a potential infinity of theorems; from finitely many words, thought can exfoliate a potential infinity of sentences; from finitely many data, reflection can extract a potential infinity of items of information. Even with respect to a world of finitely many objects, the process of reflecting upon these objects can, in principle, go on unendingly. One can inquire about their features, the features of these features, and so on. Or again, one can consider their relations, the relations among those relations, and so on. Thought – abstraction, reflection, analysis – is an inherently ampliative process. As in physical reflection mirror-images can reflect one another indefinitely, so mental reflection can go on and on. Given a start, however modest, thought can advance *ad indefinitum* into new conceptual domains. The circumstance of its starting out from a finite basis does *not* mean that it need ever run out of impetus (as the example of Shakespearean scholarship seems to illustrate).

The number of true descriptive remarks that can be made about a thing – about any actual physical object – is theoretically inexhaustible. For example, take a stone. Consider its physical features: its shape, its surface texture, its chemistry, etc. And then consider its causal background: its subsequent genesis and history. Then consider its functional aspects as relevant to its uses by the stonemason, or the architect, or the

111

landscape decorator, etc. There is, in principle, no theoretical limit to the different lines of consideration available to yield descriptive truths about a thing, so that the totality of facts about a thing – about any real thing whatever – is in principle inexhaustible.

It is helpful to introduce a distinction at this stage. If we adopt the standard conception of the matter, then a "truth" is something to be understood in *linguistic* terms – the representation of a fact through its statement in some actual language. Any correct statement in some actual language formulates a truth. (And the converse obtains as well: a *truth* must be encapsulated in a statement, and cannot exist without linguistic embodiment.) A *fact*, on the other hand, is not a linguistic entity at all, but an actual circumstance or state of affairs. Anything that is correctly statable in some *possible* language presents a fact.[1]

There are *prima facie* more facts than truths. Every truth must state a fact, but in principle it is possible that there will be facts that cannot be stated in any actually available language and thus fail to be captured as truths. Facts afford *potential* truths whose actualization hinges on the availability of appropriate linguistic apparatus for their formulation. Truths involve a one-parameter possibilization – they include whatever *can* be stated truly in some *actual* language. Facts, on the other hand, involve a two-parameter possibilization: they include whatever *can* be stated truly in some *possible* language. Truths are *actualistically* language-correlative, while facts are *possibilistically* language-correlative.[2] Accordingly, it must be presumed that there are facts which will never be formulated as truths, though it will obviously be impossible to give concrete examples of this phenomenon.[3]

Now propositional knowledge regarding matters of fact (including belief and conjecture) is always a matter of linguistically formulable information. And we have no alternative to supposing that the realm of facts regarding the real things of this world is larger than the attainable body of knowledge about them – regardless of whether that "we" is construed distributively or collectively. It is not very difficult to see why this is so.

As long as we are concerned with information formulated in languages of the standard (recursively developed) sort, the number of actually *articulated* items of information (truths or purported truths) about a thing is always, at any historical juncture, finite. And it remains *denumerably* infinite even over a theoretically endless long run.[4] The domain of truth is therefore denumerable: but that of fact is not. Our concept of the real world is such that there will always be non-denumerably many facts about a real thing.

This can be substantiated by an argument of the "diagonalization" type. For any infinite list or inventory of distinct facts about something will inevitably be such that there will always be *further* facts, not yet included, that do not occur anywhere on it. Thus, let us suppose, for the sake of *reductio ad absurdum*, that we had a non-redundantly complete enumeration of *all* of the distinct facts about something:

$$f_1, f_2, f_3, \ldots$$

Then by the supposition of *factuality* we have $(\forall i)f_i$. And by the supposition of *completeness* we have it that, as regards claims about this item:

$$(\forall p) \, [p \rightarrow (\exists i)(p \longleftrightarrow f_i)]$$

Moreover, by the supposition of *non-redundancy*, each member of the sequence adds something quite new to what has gone before.

$$(\forall i)(\forall j)(i<j \rightarrow \neg[(f_1 \, \& \, f_2 \, \& \, \ldots \, \& \, f_i) \rightarrow f_j])$$

Consider now the following course of reasoning.

(1) $(\forall i)f_i$ by "factuality"
(2) $(\exists j)[(\forall i)f_i \longleftrightarrow f_j]$ from (1) by "completeness"
(3) $(\forall i)f_i \longleftrightarrow f_j$ from (2) by existential instantiation
(4) $f_j \rightarrow f_{j+1}$ from (3) by universal instantiation
(5) $\neg[(f_1 \, \& \, f_2 \, \& \, \ldots \, \& \, f_j) \rightarrow f_{j+1}]$ from "non-redundancy" by universal instantiation
(6)   (4) and (5) are mutually contradictory.

This *reductio ad absurdum* of our hypothesis indicates that the facts about a thing are necessarily too numerous for enumeration.

It follows from these lines of thought that we cannot possibly articulate, and thus come to know explicitly, "the whole truth" about a thing. The domain of fact inevitably transcends the limits of our capacity to *express* it, and *a fortiori* those of our capacity to canvass it in overt detail. There are always bound to be more facts than we are able to capture in our linguistic terminology.

It might be possible, however, to have latent or implicit knowledge of an infinite domain through deductive systematization. After all, the finite set of axioms of a formal system will yield infinitely many theorems. And so, it might seem that when we shift from overt or explicit to implicit or tacit knowledge, we secure the prospect of capturing an infinitely diverse *implicit* knowledge-content within a finite *explicit* linguistic basis through recourse to deductive systematization.

The matter is not, however, quite so convenient. The totality of the deductive consequences that can be obtained from any finite set of axioms is itself always denumerable. The most we can ever hope to encompass by any sort of *deductively* implicit containment within a finite basis of truths is a *denumerably infinite* manifold of truths. And thus as long as implicit containment remains a recursive process, it too can never hope to transcend the range of the denumerables, and so cannot hope to encompass the whole of the transdenumerable range of descriptive facts about a thing. (Moreover, even within the denumerable realm, our attempt at deductive systematization runs into difficulties: as is known from Kurt Goedel's work, one cannot even hope to systematize – by any recursive, axiomatic process – all of the inherently denumerable truths of arithmetic. It is one of the deepest lessons of modern mathematics that we cannot take the stance that if there is a fact of the matter in this domain, then we can encompass it within the deductive means at our disposal.)

But there is yet another way of substantiating our point. For the preceding considerations related to the limits of knowledge that can be rationalized on a *fixed and given* conceptual basis – a full-formed, developed language. But, in real life, languages are never full-formed and a conceptual basis is never "fixed and given". Even with such familiar things as birds, trees, and clouds, we are involved in a constant reconceptualization in the course of progress in genetics, evolutionary theory, and thermodynamics. Our conceptions of things always present a *moving* rather than a *fixed* object of scrutiny, and this historical dimension must also be reckoned with.

Any adequate theory of inquiry must recognize that the ongoing process of information acquisition at issue in science is a process of *conceptual* innovation, which always leaves certain facts about things wholly outside the cognitive range of the inquirers of any particular period. Caesar did not know – and in the then extant state of the cognitive arts could not have known – that his sword contained tungsten and carbon. There will always be facts (or plausible candidate facts) about a thing that we do not *know* because we cannot even *conceive* of them in the prevailing order of things. To grasp such a fact means taking a perspective of consideration that as yet we simply do not have, since the state of knowledge (or purported knowledge) is not yet advanced to a point at which such a consideration is feasible. Any adequate world-view must recognize that the ongoing progress of scientific inquiry is a process of *conceptual* innovation that always leaves various facts about the things of this world wholly outside the cognitive range of the inquirers of any particular period.

The language of emergence can perhaps be deployed usefully to make the point. But what is at issue is not an emergence of *the features of things*, but an emergence in our *knowledge* about them. Blood circulated in the human body well before Harvey; substances containing uranium were radioactive before Becquerel. The emergence at issue relates to our cognitive mechanisms of conceptualization, not to the *objects* of our consideration in and of themselves. Real-world objects must be conceived of as antecedent to any cognitive interaction – as being there right along, "pregiven" as Edmund Husserl put it. Any cognitive changes or innovations are to be conceptualized as something that occurs on *our* side of the cognitive transaction, and not on the side of the *objects* with which we deal.[5]

The prospect of change can never be eliminated in this domain. The properties of a thing are literally open-ended: we can always discover more of them. Even if we were (surely mistakenly) to view the world as inherently finitistic – espousing a Keynesian principle of "limited variety" to the effect that nature can be portrayed descriptively with the materials of a finite taxonomic scheme – there will still be no *a priori* guarantee that the progress of science will not lead *ad indefinitum* to changes of mind regarding this finite register of descriptive materials. And this conforms exactly to our expectation in these matters. For where the real things of the world are concerned, we not only expect to learn more about them in the course of scientific inquiry, *we expect to have to change our minds about their nature and modes of comportment.* Be the items at issue elm trees, or volcanoes, or quarks, we have every expectation that in the course of future scientific progress people will come to think about their origin and their properties differently from the way we do at this juncture.

2. THE COGNITIVE OPACITY OF REAL THINGS

It is worthwhile to examine more closely the considerations that indicate the inherent imperfection of our knowledge of things.[6]

To begin with, it is clear that, as we standardly think about things within the conceptual framework of our fact-oriented thought and discourse, *any* real physical object has more facets than it will ever actually manifest in experience. For every objective property of a real thing has consequences of a dispositional character and these are never surveyable *in toto* because the dispositions which particular concrete things inevitably have endow them with an infinitistic aspect that cannot be comprehended within experience.[7] This desk, for example, has a limitless manifold of phenomenal features of the type: "having a certain appearance from a particular point of view". It is perfectly clear that most of these will never be actualized in experience. Moreover, a thing *is* what it *does*: entity and lawfulness are coordinated correlates – a good Kantian point. And this fact that things demand lawful comportment means that the finitude of experience precludes any prospect of the *exhaustive* manifestation of the descriptive facets of any real things.[8]

Physical things not only have more properties than they ever will overtly manifest, but they have more than they can possibly ever actually manifest. This is so because the dispositional properties of things always involve what might be characterized as *mutually preemptive* conditions of realization. This cube of sugar, for example, has the dispositional property of reacting in a particular way if subjected to a temperature of 10 000°C and of reacting in a certain way if emplaced for one hundred hours in a large, turbulent body of water. But if either of these conditions is ever realized, it will destroy the lump of sugar as a lump of sugar, and thus block the prospect of *its* ever bringing the other property to manifestation. The perfectly possible realization of various dispositions may fail to be mutually *compossible*, and so the dispositional properties of a thing cannot ever be manifested completely – not just in practice, but in principle. Our objective claims about real things always commit us to more than we can actually ever determine about them.

The existence of this latent (hidden, occult) sector is a crucial feature of our conception of a real thing. Neither in fact nor in thought can we ever simply put it away. To say of the apple that its only features are those it actually manifests is to run afoul of our conception of an apple.

To deny – or even merely to refuse to be committed to the claim – that it *would* manifest particular features *if* certain conditions came about (for example, that it would have such-and-such a taste if eaten) is to be driven to withdrawing the claim that it is an apple. The process of corroborating the implicit contents of our objective factual claims about something real is potentially endless, and such judgments are thus "non-terminating", in C. I. Lewis' sense.[9] This cognitive depth of our objective factual claims inherent in the fact that their *content* will always outrun the evidence for making them means that the endorsement of any such claim always involves some element of evidence-transcending conjecture.

The concepts at issue (viz. "experience" and "manifestation") are such that we can only ever *experience* those features of a real thing that it actually *manifests*. But the preceding considerations show that real things always have more experientially manifestable properties than they can ever actually manifest in experience. The experienced portion of a thing is similar to the part of the iceberg that shows above water. All real things are necessarily thought of as having hidden depths that extend beyond the limits, not only of experience, but also of experientiability. To say of something that it is an apple or a stone or a tree is to become committed to claims about it that go beyond the data we have – and even beyond those which we can, in the nature of things, ever actually acquire. The "meaning" inherent in the assertoric commitments of our factual statements is never exhausted by its verification. Real things are cognitively opaque – we cannot see to the bottom of them. Our knowledge of such things can thus become more *extensive* without thereby becoming more *complete*.

In this regard, however, real things differ in an interesting and important way from their fictional cousins. To make this difference plain, it is useful to distinguish between two types of information about a thing, namely that which is *generic* and that which is not. Generic information tells about those features of a thing which it has in common with everything else of its kind or type. For example, a particular snowflake will share with all others certain facts about its structure, its hexagonal form, its chemical composition, its melting point, etc. On the other hand, it will also have various properties which it does not share with other members of its own "lowest species" in the classificatory order – its particular shape, for example, or the angular momentum of its descent. These are its non-generic features.

Now a key fact about *fictional* particulars is that they are of finite cognitive depth. In discoursing about them we shall ultimately run out of steam as regards their non-generic features. A point will always be reached when one cannot say anything further that is characteristically new about them – presenting non-generic information that is not inferentially implicit in what has already been said. New *generic* information can, of course, always be forthcoming through the progress of science. When we learn more about coal-in-general then we know more about the coal in Sherlock Holmes' grate. But the finiteness of their cognitive depth means that the presentation of ampliatively novel *non-generic* information must by the very nature of the case come to a stop when fictional things are at issue.

With *real* things, on the other hand, there is no reason of principle why the provision of non-generically idiosyncratic information need ever be terminated. On the contrary, we have every reason to presume these things to be cognitively inexhaustible. A precommitment to description-transcending features – no matter how far description is pushed – is essential to our conception of a real thing. Something whose character was exhaustible by linguistic characterization would thereby be marked as fictional rather than real.[10]

This cognitive opacity of real things means that we are not – and will never be – in a position to evade or abolish the contrast between "things as we think them to be" and "things as they actually and truly are". Their susceptibility to further elaborative detail – and to changes of mind regarding this further detail – is built into our very conception of a "real thing". To be a real thing is to be something regarding which we can always, in principle, acquire more and possibly discordant information. This view of the situation is supported rather than impeded once we abandon the naive cumulativist/preservationist view of knowledge acquisition for the view that new discoveries need not *supplement* but can *displace* old ones. We realize that people will come to think differently about things from the way we do – even when thoroughly familiar things are at issue – recognizing that scientific progress generally entails fundamental changes of mind about how things work in the world.

In view of the cognitive opacity of real things, we must never pretend to a cognitive monopoly or cognitive finality. This recognition of incomplete information is inherent in the very nature of our conception of a "real thing". It is a crucial facet of our epistemic stance towards the real world to recognize that every part and parcel of it has features lying beyond our present cognitive reach – at *any* "present" whatsoever.

Much the same story holds when our concern is not with physical things, but with *types* of such things. To say that something is copper or magnetic is to say more than that it has the properties we think copper or magnetic things have, and to say more than that it meets our test conditions for being copper (or being magnetic). It is to say that this thing *is* copper or magnetic. And this is an issue regarding which we are prepared at least to contemplate the prospect that we have got it wrong.

Certainly, it is imaginable that natural science will come to a stop, not in the trivial sense of a cessation of intelligent life, but in Charles Sanders Peirce's more interesting sense of eventually reaching a condition after which even indefinitely ongoing effort at inquiry will not – and indeed actually *cannot* – produce any significant change. Such a position is, in theory, possible. But we can never *know* – be it in practice or in principle – that it is actually realized. We can never establish that science has attained such an omega-condition of final completion: the possibility of further change lying "just around the corner" can never be ruled out finally and decisively. Thus, we have no alternative but to *presume* that our science is still imperfect and incomplete, that no matter how far we have pushed our inquiries in any direction, regions of *terra incognita* yet lie beyond.

3. THE CORRIGIBILITY OF CONCEPTIONS

It must be stressed that these deliberations regarding cognitive inadequacy are less concerned with the correctness of our *particular claims* about real things than with our *characterizing conceptions* of them. And in this connection it deserves stressing that there is a significant and substantial difference between a true or correct *statement* or *contention* on the one hand, and a true or correct *conception* on the other. To make a true contention about a thing we merely need to get *one particular fact* about it straight. To have a true conception of the thing, on the other hand, we must get *all of the important facts* about it straight. And it is clear that this involves a certain *normative* element – namely what the "important" or "essential" facets of something are.

Anaximander of Miletus presumably made many correct contentions about the sun in the fifth century B.C. – for example, that its light is brighter than that of the moon. But Anaximander's *conception* of the sun (as the flaming spoke of a great wheel of fire encircling the earth) was totally wrong.

To assure the correctness of our conception of a thing we would have to be sure – as we very seldom are – that nothing further can possibly come along to upset our view of just what its important features are and just what their character is. Thus, the qualifying conditions for true conceptions are far more demanding than those for true claims. With a correct contention about a thing, all is well if we get the single relevant aspect of it right, but with a correct conception of it *we must get the essentials right* – we must have an overall picture that is basically correct. And this is something we generally cannot ascertain, if only because we cannot say with confidence what is really important or essential before the end of the proverbial day.

With *conceptions* – unlike propositions or *contentions* – incompleteness means incorrectness, or at any rate *presumptive* incorrectness. Having a correct or adequate conception of something as the object it is requires that we have all the *important* facts about it right. But since the prospect of discovering further important facts can never be eliminated, the possibility can never be eliminated that matters may so eventuate that we may ultimately (with the wisdom of hindsight) acknowledge the insufficiency or even inappropriateness of our earlier conceptions. A conception based on incomplete data must be assumed to be at least partially incorrect. If we can decipher only half an inscription, our conception of its overall content must be largely conjectural – and thus must be *presumed* to contain an admixture of error. When our information about something is incomplete, obtaining an overall picture of the thing at issue becomes a matter of theorizing, or guesswork, however sophisticatedly executed. And then we have no alternative but to suppose that this overall picture falls short of being wholly correct in various (unspecifiable) ways. With conceptions, falsity can thus emerge from errors of *omission* as well as those of *commission*, resulting from the circumstance that the information at our disposal is merely incomplete, rather than actually false (as would have to be the case with contentions).

To be sure, an inadequate or incomplete *description* of something is not thereby false – the statements we make about it may be perfectly true as far as they go. But an inadequate or incomplete *conception* of a thing is *ipso facto* one that we have no choice but to presume to be *incorrect* as well,[11] seeing that where there is incompleteness we cannot justifiably take the stance that it relates only to inconsequential matters and touches nothing important. Accordingly, our conceptions of particular things are always to be viewed not just as cognitively *open-ended*, but as *corrigible* as well.

We are led back to the thesis of the great idealist philosophers (Spinoza, Hegel, Bradley, Royce) that human knowledge inevitably falls short of "perfected science" (the Idea, the Absolute), and must be presumed deficient both in its completeness and its correctness.[12]

4. PERSPECTIVES ON REALISM

Let us now return to our central theme of realism. Physical realism claims the objective reality of physical existents. But what is involved in such "objectivity"? What are we committing ourselves to in saying of something that it is "a real thing", an object existing as part of the world's furniture? Clearly, we commit ourselves to several (obviously interrelated) points.

(1)   *Substantiality of entity*. Being a "something" (entity) with its own unity of being. Having an enduring identity of its own.

(2)   *Physicality or reality*. Existing in space and time. Having a place as a real item in the world's physical scheme.

(3)   *Publicity or accessibility*. Admitting universality of access. Being something that different investigators proceeding from different points of departure can get hold of.

(4)   *Autonomy or independence*. Being independent of mind. Being something that observers find rather than create.

In natural science we try to get at the objective matters of fact regarding physical reality in ways that are accessible to all observers alike. (The "repeatability of experiments" is crucial.) And the salient factor enters in with that fourth and final issue – autonomy. The very idea of a thing so functions in our conceptual scheme that real things are thought of as having an identity, a nature, and a mode of comportment wholly indifferent to the cognitive state-of-the-art regarding them – and potentially even very different from our own current conceptions of the matter.

The conception of a *thing* that underlies our discourse about the things of this world reflects a certain sort of tentativity and fallibilism – the implicit recognition that our own personal or even communal conception of particular things may, in general, be wrong, and is in any case inadequate. At the back of our thought about things there is always a certain wary scepticism that recognizes the possibility of error. The objectivity of real existents projects beyond the reaches of our subjectively conditioned information.

There is wisdom in Hamlet's dictum: "There are more things in heaven and on earth, Horatio . . ." The limits of our knowledge may be the limits of *our* world, but they are not the limits of *the* world. We do and must recognize the limitations of our cognition. We cannot justifiably equate reality with what can, in principle, be known by us, nor equate reality with what can, in principle, be expressed by our language. And what is true here for our sort of mind is true for any other sort of finite mind as well. Any physically realizable sort of cognizing being can only know a part or aspect of the real.

The issue of "objectivity" in the sense of mind-independence is pivotal for realism. A fact is objective in this mode if it obtains independently of whatever thinkers may think about relevant issues, so that changes merely in what is thought by the world's intelligences would leave it unaffected – that is, if it is thought-invariant or thought-insensitive. With objective facts (unlike those which are merely a matter of inter-subjective agreement) what thinkers think just does not enter into it. What realism maintains from the outset – and traditional idealism often struggles valiantly to retain (with mixed success) – is just this conception that there are certain objective facts that obtain to things independently, regardless of what we, or anybody else, think of them.

At this point we reach an important conjuncture of ideas. The ontological independence of things – their objectivity and autonomy of the machinations of mind – is a crucial aspect of realism. And the fact that it lies at the very core of our conception of a real thing that such items project beyond the cognitive reach of mind betokens a concept-scheme fundamentally committed to objectivity. The only plausible sort of ontology is one that contemplates a realm of reality that outruns the range of knowledge (and indeed even of language), adopting the stance that character goes beyond the limits of characterization. It is a salient aspect of the mind-independent status of the objectively real that the features of something real always transcend what we know about it. Indeed, yet further or different facts concerning a real thing can always come to light, and all that we *do* say about it does not exhaust all that *can and should* be said about it. In this light, objectivity is crucial to realism and the cognitive inexhaustibility of things is a certain token of their objectivity.

Authentic realism can only exist in a state of tension. The only reality worth having is one that is in some degree knowable. But it is the very limitation of our knowledge – our recognition that there is more to reality than what we do and can know or ever conjecture about it – that speaks for the mind-independence of the real. It is important to stress against the sceptic that the human mind is sufficiently well attuned to reality that *some* knowledge of it is possible. But it is no less important to join with realists in stressing the independent character of reality, acknowledging that reality has a depth and complexity of make-up that outruns the reach of mind.

Peirce and others have located the impetus to realism in the limitations of man's will – in the fact that we can exert no control over our experience and, try as we will, cannot affect what we see and sense. Peirce's celebrated "Harvard Experiment" of the Lowell Lectures of 1903 makes the point forcibly.

I know that this stone will fall if it is let go, because experience has convinced me that objects of this kind always do fall; and if anyone present has any doubt on the subject, I should be happy to try the experiment, and I will bet him a hundred to one on the result . . . [I know this because of an unshakable conviction that] the uniformity with which stones have fallen has been due to some *active general principle* [of nature] . . .

Of course, every sane man will adopt the latter hypothesis. If he could doubt it in the case of the stone – which he can't – and I may as well drop the stone once and for all – I told you so! – if anybody doubt this still, a thousand other such inductive predictions are getting verified every day, and he will have to suppose every one of them to be merely fortuitous in order reasonably to escape the conclusion that *general principles are really operative in nature*. That is the doctrine of scholastic realism.[13]

No doubt, the ordinary man, and most philosophers with him, stands committed to the conviction that whatever happens in the world of observation happens in line with the causally lawful machinations of an underlying mind-independent physical reality.

In this context, however, it is important to distinguish between mental *dependency* and mental *control*. Peirce is clearly right in saying that we cannot *control* our conviction that the stone will fall: do what we will, it will remain. Nevertheless, this circumstance could conceivably still be something that *depends on us* – exactly as with the fearsomeness of heights for the man with vertigo. If the *unconscious* sphere of mind actually dictates how I *must* "see" something (as, for example, in an optical illusion of the Mueller–Leyer variety) then I evidently have no *control*. But that does not *in itself* refute mind-dependency – even of a very strong sort. There is always the prospect that we are deluding ourselves in these matters – that the limitations at issue appertain only to our *conscious* powers, and not to our powers as such.

This prospect blocks Peirce's argument in the way already foreseen by Descartes in the *Meditations*.

I found by experience that these [sensory] ideas presented themselves to me without my consent being requisite, so that I could not perceive any object, however desirous I might be, unless it were present . . . But although the ideas which I receive by the senses do not depend on my will I do not think that one should for that reason conclude that they proceed from things different from myself, since possibly some facility might be discovered in me – though different from those yet known to me – which produced them.[14]

We may simply delude ourselves about the range of the mind's powers: lack of control notwithstanding, dependency may yet lie with the "unconscious" sector of mind. The traditional case for realism based on the limits of causal control thus fails to provide a really powerful argument for mind-independence.

However, a far more effective impetus to realism lies in the limitations of man's *intellect*, pivoting on the circumstance that the features of real things inevitably outrun our cognitive reach. In placing some crucial aspects of the real altogether outside the effective range of mind, it speaks for a position that sees mind-independence as a salient feature of the real. The very fact of fallibilism and limitedness – of our absolute confidence that our putative knowledge does *not* do justice to the real truth of the matter of what reality is actually like – is surely one of the best arguments for a realism that pivots on the basic idea that there is more to reality than we humans do or can know about. Traditional scientific realists see the basis for realism in the substantive knowledge of the sciences; the present realism, by contrast, sees its basis in our realization of the inevitable *shortcomings* of our scientific knowledge.

This line of approach preempts the preceding sort of objection. If we are mistaken about the reach of our cognitive powers – if they do not adequately grasp "the way things really are" – then this very circumstance clearly *bolsters* the case for realism. The cognitive intractability of things is something about which, in principle, we cannot delude ourselves, since such delusion would illustrate rather than abrogate the fact of a reality independent of ourselves. The very inadequacy of our knowledge is one of the most salient tokens there is of a reality out there that lies beyond the inadequate gropings of mind. It is the very limitation of our knowledge of things – our recognition that reality extends beyond the horizons of what we can possibly know or even conjecture about it – that betokens the mind-independence of the real.

A brief postscript is in order here. One must be careful about what the presently contemplated sort of argument for realism actually manages to establish. For it does *not* establish outright that a stone – be it Peirce's or Dr. Johnson's or the geologist's – is something mind-independently real. Rather, what it shows is *that our conception of a "stone" – our conception of a physical object – is the conception of something that is mind-independently real*. And so the realism underwritten by these deliberations is not in fact a squarely *ontological* doctrine, but a realism geared to our conceptual scheme for thinking about things. It is a certain sort of radical idealism that is the target of these deliberations; radical scepticism is the topic for another occasion.[15]

CHAPTER TEN

METAPHYSICAL REALISM AND THE PRAGMATIC BASIS OF OBJECTIVITY

SYNOPSIS. (1) The ontological component of metaphysical realism – the idea that there indeed is an objective sphere of mind-independent reality which exists independently in its own right, without reference to anyone's ideas and conceptions about it – is not something that we learn from experience. It is a *presupposition* for our experience-exploiting inquiries, rather than a product thereof. (2) We have to do here with a postulation made on *functional* rather than *evidential* grounds, which we endorse in order to be in a position to learn by experience at all. This postulation is justified in the first analysis on the grounds of functional requiredness, seeing that it is an indispensable requisite for our standard conceptual scheme with respect to inquiry, cognition, and discourse. (3) The project of communal inquiry into and inter-personal communication about an objective order of reality plays an especially important justificatory role. (4) Moreover, without the resource of an objective order of impersonal fact, we would be thrown back on totally non-cognitive means for the guidance of action. (5) The validation of the reality postulate thus lies *originally* in its potential functional utility, and *ultimately* in its being retro-justified by the "wisdom of hindsight" on grounds of its pragmatic and explanatory efficacy.

1. THE EXISTENTIAL COMPONENT OF REALISM

Realism has two indispensable and inseparable components – the one existential and ontological, and the other cognitive and epistemic. The former maintains that there indeed is a real world – a realm of mind-independent, objective physical reality. The latter maintains that we can to some extent secure adequate information about this mind-independent realm. This second contention obviously presupposes the first. But how can that first, ontological thesis be secured?

Metaphysical realism is clearly not an inductive inference secured through the scientific systematization of our observations, but rather a regulative presupposition that makes science possible in the first place. The realm of mind-independent reality is something we cannot *discover* – we do not learn that it exists as a fruit of inquiry and investigation. How could we ever learn by inference from observations that our observations are objectively valid, that our mental experience is itself largely the causal product of the machinations of a mind-independent matrix, that all those phenomenal appearances are causally rooted in a physical reality? All this is clearly something we do not *learn* from inquiry. For what is at issue is, after all, a *precondition* for empirical inquiry – a presupposition for the usability of observational data as sources of objective information. That experience is indeed objective, that what we take to be evidence *is* evidence, that our sensations yield information about an order of existence outside the experiential realm itself, and that this experience constitutes not just a mere phenomenon but an appearance of something extra-mental belonging to an objectively self-subsisting order. All this is something that we must always *presuppose* in using experiential data as "evidence" for how things stand in the world. Objectivity represents a postulation made on *functional* rather than *evidential* grounds: we endorse it in order to be in a position to learn by experience at all. We do not learn or discover that there is a mind-independent physical reality, we *presume or postulate* it. As Kant clearly saw, objective experience is possible only if the existence of such a real, objective world is *presupposed* from the outset rather than seen as a matter of *ex post facto* discovery about the nature of things.[1]

Accordingly, the crucial existential (ontological) component of realism is not a matter of discovery, a part of the findings of empirical research. It is a presupposition for our inquiries rather than a result thereof. We have to do here not with an *evidentiated discovery* about the constitution of nature as such, but rather with a *formative assumption* that undergirds our view of the nature of inquiry. Without subscribing to this idea, we could not think of our knowledge as we actually do. Our commitment to the existence of a mind-independent reality is thus a postulate whose justification pivots on its functional utility in enabling us to operate as we do with respect to inquiry.

Of course, after we postulate an objective reality and its concomitant causal aspect, then principles of inductive systematization, of explanatory economy, and of common cause consilience can work wonders towards furnishing us with plausible claims about the nature of the real. But we indispensably need that initial existential presupposition to make a start. Without commitment to a reality to serve as ground and object of our experience, its cognitive import will be lost. Only on this basis can we proceed evidentially with the exploration of the interpersonally public and objective domain of a physical world-order that we share in common.

To be sure, that second, descriptive (epistemic) component of realism stands on a very different footing. Reality's *nature* is something about which we can only make warranted claims through examining it. Substantive information must come through inquiry – through evidential validation. Once we are willing to credit our observational data with objectivity, and thus with evidential bearing, we can, of course, make use of them to inform ourselves as to the nature of the real.

Our endorsement of unobserved causes in nature is not based on science but on metaphysics. What we learn from science is not *that* a (thus far) unobservable order of physical existence causally undergirds nature as we observe it, but rather *what* these underlying structures are like. Science does not (cannot) teach us that the observable order is explicable in terms of underlying causes and that the phenomena of observation are signs or symptoms of this extra- and sub-phenomenal order of existence; we know this *a priori* of any world in which observation as we understand it can transpire. What science does teach us (and metaphysics cannot) is what the descriptive character of this extra-phenomenal order is in the context of our world.

Let us consider this basic reality postulate somewhat more closely. Our standard conception of inquiry involves recognition of the following facts. (1) The world (the realm of physical existence) has a nature whose characterization in point of description, explanation, and prediction is the object of empirical inquiry. (2) The real nature of the world is in the main independent of the process of inquiry which the real world canalizes or conditions. Dependency is a one-way street here: reality shapes or influences inquiry, but not conversely. Our opinions do not affect the real truth but, rather, our strivings after the real truth engender changes in our opinions. (3) In virtue of these considerations, we can stake neither total nor final claims for our purported knowledge of reality. Our knowledge of the world must be presumed incomplete, incorrect, and imperfect, with the consequence that "our reality" must be considered to afford an inadequate characterization of "reality itself".

The crucial question is this: Assuming that there *are* objective facts, how can we possibly come to acquire knowledge of them? What sort of presuppositions must we make if our subjective experience – which is limited and episodic – is to provide a basis of legitimacy for maintaining objective and general claims?

Two gulfs must be transcended:

(1)   From experiential appearances to objective facts – from

"That looks like a red apple to me" to "That is a red apple" –
from phenomena to real things.

(2)   From particular cases to universals – from "This apple contains seeds" to "All apples contain seeds".

To effect these transitions, we must simply *presuppose* (for how could
we possibly *preestablish* this?) that these moves can be made on the
basis of available evidence: that subjective phenomena are indicators
from objective realities and that particular cases are exemplifications of
universal arrangements.

The foundations of objectivity do not rest on the findings of science.
They precede and underlie science, which would itself not be possible
without a precommitment to the capacity of our senses to warrant claims
about an objective world order. Objectivity is not a *product* of inquiry;
we must precommit ourselves to it to make inquiry possible. It is a
necessary input into the cognitive project and not a contingent output
thereof. The objective bearing of experience is not something we can
preestablish; it is something we must presuppose in the interest of
honoring Peirce's cogent injunction never to bar the path of inquiry.

With respect to our cognitive endeavors, "man proposes and nature
disposes", and it does so in both senses of the term: it disposes *over* our
current view of reality and it will doubtless eventually dispose *of* it as
well. Our view of the nature of inquiry and of the sort of process it
represents is possible only because we stand committed from the very
outset to the idea of ourselves as a minuscule component of a mind-
independent reality. We can act and affect a few things in it. But in the
main it has the whip hand and we merely respond to its causal dictates.
And this is true in cognitive aspects as well – where we must simply do
the best we can with the relatively feeble means at our disposal.

Our. commitment to realism pivots on a certain practical *modus
operandi*, encapsulated in the precept: "Proceed in matters of inquiry
and communication on the basis that you are dealing with an objective
realm, existing quite independently of the doings and dealings of
minds." Accordingly, we standardly operate on the basis of the "presumption of objectivity" reflected in the guiding precept: "Unless you
have good reason to think otherwise (that is, as long as *nihil obstat*) treat
the materials of inquiry and communication as veridical." The ideal of
objective reality is the focus of a family of convenient regulative prin-

ciples – a functionally useful instrumentality that enables us to transact our cognitive business in the most satisfactory and rewarding way.

2. REALISM IN ITS REGULATIVE/PRAGMATIC ASPECT

What legitimates metaphysical realism's postulation that experience affords data regarding an objective and mind-independent domain and thus provides for viable information about the real? Given that the existence of such a domain is not a product of empirical inquiry but a precondition for it, its acceptance has to be validated in the manner appropriate for postulates and prejudgments of any sort – namely in terms of its prospective utility. Bearing this pragmatic perspective in mind, let us take a closer look at this issue of utility and ask: What can postulation of a mind-independent reality actually do for us?

The answer is straightforward. The assumption of a mind-independent reality is essential to the whole of our standard conceptual scheme relating to inquiry and communication. Without it, both the actual conduct and the rational legitimation of our communicative and investigative (evidential) practice would be destroyed. Nothing that we do in this cognitive domain would make sense if we did not subscribe to the conception of a mind-independent reality.

To begin with, we indispensably require the notion of reality to operate the classical concept of truth as "agreement with reality" (*adaequatio ad rem*). Once we abandon the concept of reality, the idea that in accepting a factual claim as true we become committed to how matters actually stand – "how it really is " – would also go by the board. Semantics constrain realism; we have no alternative but to regard as real those states of affairs claimed by the contentions we are prepared to accept. Once we put a contention forward by way of serious assertion, we must view as real the states of affairs it purports, and must see its claims as facts. We need the notion of reality to operate the conception of truth. A factual statement on the order of "There are pi mesons" is true if and only if the world is such that pi mesons exist within it. By virtue of their very nature as truths, true statements must state facts; they state what really is so, which is exactly what it is to "characterize reality". The conceptions of *truth* and of *reality* come together in the notion of *adaequatio ad rem* – the venerable principle that to speak truly is to say how matters stand in reality, in that things actually are as we have said them to be.

In the second place, the nihilistic denial that there is such a thing as reality would destroy once and for all the crucial Parmenidean divide between appearance and reality. And this would exact a fearful price from us: we would be reduced to talking only of what we *think* to be so. The crucial contrast notion of the *real* truth would no longer be available: we would only be able to contrast our *putative* truths with those of others, but could no longer operate the classical distinction between the putative and the actual, between what we think to be so and what actually *is* so. We could not take the stance that, as the Aristotelian commentator Themistius put it, "that which exists does not conform to various opinions, but rather the correct opinions conform to that which exists."[2]

The third point is the issue of cognitive coordination. Communication and inquiry, as we actually carry them on, are predicated on the fundamental idea of a real world of objective things, existing and functioning "in themselves", without dependence on us and so equally accessible to others. Inter-subjectively valid communication can only be based on common access to an objective order of things. All our ventures at communication and communal inquiry are predicated on the stance that we communally inhabit a shared world of objectively existing things. There is a realm of "real objects" amongst which we live and into which we inquire as a community, but about which we ourselves as individuals presumably have only imperfect information that can be criticized and augmented by the efforts of others.

This points to a fourth important consideration. Only through reference to the real world as a *common object* and shared focus of our diverse and imperfect epistemic strivings are we able to effect communicative contact with one another. Inquiry and communication alike are geared to the conception of an objective world: a communally shared realm of things that exist strictly "on their own", comprising an enduring and independent realm within which and, more importantly, with reference to which inquiry proceeds. We could not operate the notion that inquiry estimates the character of the real if we were not prepared to presume or postulate a reality for these estimates to be estimates of. It would clearly be pointless to devise our characterizations of reality if we did not stand committed to the proposition that there is a reality to be characterized.

The fifth item is a recourse to mind-independent reality which makes possible a "realistic" view of our knowledge as potentially flawed. A rejection of this commitment to reality *an sich* (or to the actual truth about it) exacts an unacceptable price. For in abandoning this commitment we also lose those regulative contrasts that canalize and condition our view of the nature of inquiry (and indeed shape our conception of this process as it stands within the framework of our conceptual scheme). We could no longer assert: "What we have there is good enough as far as it goes, but it is presumably not 'the whole real truth' of the matter." The very conception of inquiry as we conceive it would have to be abandoned if the contrast conceptions of "actual reality" and "the real truth" were no longer available. Without the conception of reality we could not think of our knowledge in the fallibilistic mode we actually use – as having provisional, tentative, improvable features that constitute a crucial part of the conceptual scheme within whose orbit we operate our concept of inquiry.

Reality (on the traditional metaphysicians' construction of the concept) is the condition of things answering to "the real truth"; it is the realm of what really is as it really is. The pivotal contrast is between "mere appearance" and "reality as such", between "our picture of reality" and "reality itself", between what actually is and what we merely think (believe, suppose, etc.) to be. And our allegiance to the conception of reality, and to this contrast that pivots upon it, roots in the fallibilistic recognition that, at the level of the detailed specifics of scientific theory, anything we presently hold to be the case may well turn out otherwise – indeed, certainly will do so if past experience gives any auguries for the future.

Our commitment to the mind-independent reality of "the real world" stands together with our acknowledge that, in principle, any or all of our *present* scientific ideas as to how things work in the world, at *any* present, may well prove to be untenable. Our conviction in a reality that lies beyond our imperfect understanding of it (in all the various senses of "lying beyond") roots in our sense of the imperfections of our scientific world-picture – its tentativity and potential fallibility. In abandoning our commitment to a mind-independent reality, we would lose the impetus of inquiry.

Sixthly and finally, we need the conception of reality in order to operate the causal model of inquiry about the real world. Our standard picture of man's place in the scheme of things is predicated on the fundamental idea that there is a real world (however imperfectly our inquiry may characterize it to be) whose causal operations produce *inter alia* causal impacts upon us, providing the basis of our world-picture. Reality is viewed as the causal source and basis of the appearances, the originator and determiner of the phenomena of our cognitively relevant experience. "The real world" is seen as causally operative both in serving as the external moulder of thought and as constituting the ultimate arbiter of the adequacy of our theorizing. (Think here again of C. S. Peirce's "Harvard experiment".)

In summary, then, we need that postulate of an objective order of mind-independent reality for at least six important reasons.

(1)   To preserve the distinction between true and false with respect to factual matters and to operate the idea of truth as agreement with reality.

(2)   To preserve the distinction between appearance and reality, between our *picture* of reality and reality itself.

(3)   To serve as a basis for inter-subjective communication.

(4)   To furnish the basis for a shared project of communal inquiry.

(5)   To provide for the fallibilistic view of human knowledge.

(6)   To sustain the causal mode of learning and inquiry and to serve as basis for the objectivity of experience.

The conception of a mind-independent reality plays a central and indispensable role in our thinking with respect to matters of language and cognition. In communication and inquiry alike we seek to offer answers to our questions about how matters stand in this "objective realm". It is seen as the epistemological *object* of veridical cognition, in the context of the contrast between "the real" and its "merely phenomenal" appearances. Again, it is seen as the target or *telos* of the truth-estimation process at issue in inquiry, providing for a common focus in communication and communal inquiry. (The "real world" thus constitutes the "object" of our cognitive endeavors in both senses of this term – the *objective* at which they are directed and the *purpose* for which they are exerted.) And further, reality is seen as the ontological *source* of cognitive endeavors, affording the existential matrix in which we move and have our being, and whose impact upon us is the prime mover for our cognitive efforts. All of these facets of the concept of reality are integrated and unified in the classical doctrine of truth as it corresponds to fact (*adaequatio ad rem*), a doctrine that only makes sense in the setting of a commitment to mind-independent reality.

Accordingly, the justification for this fundamental presupposition of objectivity is not *evidential* at all; postulates are not based on evidence. Rather, it is *functional*. We need this postulate to operate our conceptual scheme. The justification of this postulate lies in its utility. We could not form our existing conceptions of truth, fact, inquiry, and communication without presupposing the independent reality of an external world. We simply could not think of experience and inquiry as we do. (What we have here is a "transcendental argument" from the character of our conceptual scheme to its inherent presuppositions.) The primary validation of that crucial objectivity postulate lies in its basic functional utility in relation to our cognitive aims. Let us explore this aspect of the matter more fully.

3. OBJECTIVITY AS A REQUISITE OF COMMUNICATION AND INQUIRY

The information that we may have about a thing, be it real or presumptive information, is always just that – information *we* lay claim to. We recognize that it varies from person to person. Our attempts at communication and inquiry are thus undergirded by an information-transcending stance – the stance that we communally inhabit a shared world of objectively existing things, a world of "real things" amongst which we live and into which we inquire (but about which we do and must presume ourselves to have only imperfect information at any and every particular stage of the cognitive venture). This is clearly not something that we learn from the course of experience. The "facts of experience" can never reveal it to us. It is something we postulate or presuppose. Its epistemic status is not that of an empirical discovery, but that of a presupposition whose ultimate justification is a transcendental argument from the very possibility of the projects of communication and inquiry as we standardly conduct them.

Our commitment to an objective reality that lies behind the data at hand is indispensably demanded by any step into the domain of the publicly accessible objects essential to communal inquiry and interpersonal communication about a shared world. We could not establish communicative contact about a common objective item of discussion if our discourse were geared to the substance of our own idiosyncratic ideas and conceptions. But the objectivity at issue in our communicative discourse is a matter of its *status* rather than one of its *content*. For the substantive content of a claim about the world in no way tells us whether it is factual or fictional. This is something we have to determine from its *context*: it is a matter of the frame, not of the canvas. The fact-oriented basis of our information-transmitting exchanges is provided *a priori* by a conventionalized intention to talk about "the real world". This intention to take real objects to be at issue, objects as they really are, our potentially idiosyncratic conceptions of them quite aside, is fundamental because it is overriding – that is, it overrides all of our other intentions when we enter upon the communicative venture. Without this conventionalized intention we should not be able to convey information – or misinformation – to one another about a shared "objective" world.

We are able to say something about the (real) moon or the (real) Sphinx because of our submission to a fundamental communicative convention or "social contract" to the effect that we *intend* ("mean") to talk about it – that very thing itself as it "really" is – our own private conception of it notwithstanding. We arrive at the standard policy that prevails with respect to all communicative discourse of letting "the language we use", rather than whatever specific informative aims we may actually "have in mind" on particular occasions, be the decisive factor with regard to the things at issue in our discourse. When I speak about the Sphinx (even though I do so on the basis of my own conceivably strange conception of what is involved here), I will be discussing "the *real* Sphinx" in virtue of the basic conventionalized intention governing our use of referring terms.

This fundamental intention of objectification, the intention to discuss "the moon itself" regardless of how untenable one's own *ideas* about it may eventually prove to be is a basic precondition of the very possibility of communication. It is crucial to the communicative enterprise to take the egocentrism-avoiding stance that rejects all claims to a privileged status for *our own* conception of things. In the interests of this stance we are prepared to "discount any misconceptions" (our own included) about things over a very wide range indeed – that we are committed to the stance that factual disagreements as to the character of things are communicatively irrelevant within very broad limits. The incorrectness of conceptions is venial.

If we were to set up our own conception of things as somehow definitive and decisive, we would at once erect a barrier not only to further inquiry but – no less importantly – to the prospect of successful communication with one another. Communication could then only proceed with the wisdom of hindsight – at the end of a long process of tentative checks. Communicative contact would be realized only in the implausible case where extensive exchange indicated retrospectively that there had been an *identity* of conceptions all along. And we would always stand on very shaky ground. For no matter how far we push our investigation into the issue of an identity of conceptions, the prospect of a divergence lying just around the corner – waiting to be discovered if only we pursued the matter just a bit further – can never be precluded. One could never advance the issue of the identity of focus past the status of a more or less well-grounded *assumption*. And then any so-called communication would no longer be an exchange of information but a tissue of frail conjectures. The communicative enterprise would become a vast inductive project – a complex exercise in theory-building, leading tentatively and provisionally toward something which, in fact, the imputational groundwork of our language enables us to presuppose from the very outset.[3]

Communication requires not only common *concepts* but common *topics*, inter-personally shared items of discussion, a common world constituted by the self-subsistently real objects basic to shared experience. The factor of objectivity reflects our basic commitment to a communally available world as the common property of communicators. Such a commitment involves more than merely *de facto* inter-subjective agreement. For such agreement is a matter of *a posteriori* discovery, while our view of the nature of things puts "the real world" on a necessary and *a priori* basis. This stance roots in the fundamental convention of a shared social insistence on communicating – the commitment to an objective world of real things affords the crucially requisite common focus needed for any genuine communication. What links my discourse with that of my interlocutor is our common subscription to the *a priori* presumption (a defeasible presumption, to be sure) that we are both talking about a shared thing, our own possible misconceptions of it notwithstanding. This means that no matter how extensively we may change our minds about the *nature* of a thing or type of thing, we are still dealing with exactly the same thing or sort of thing. It assures reidentification across theories and belief-systems.

Our concept of a *real thing* is such that it provides a fixed point, a stable center around which communication revolves, an invariant focus of potentially diverse conceptions. What is to be determinative, decisive, definitive, etc., of the things at issue in my discourse is not my conception, or yours, or indeed anyone's conception at all. The conventionalized intention discussed means that a coordination of conceptions is not decisive for the possibility of communication. Your statements about a thing will convey something to me even if my conception of it is altogether different from yours. To communicate we need not take ourselves to share views of the word, but only take the stance that we share the world being discussed.

The commitment to *objectivity* is basic to any prospect of our discourse with one another about a shared world of "real things", to which none of us is in a position to claim privileged access. This commitment establishes a need to "distance" ourselves from things, that is, to recognize the prospect of a discrepancy between our (potentially idiosyncratic) conceptions of things and the true character of these things as they exist objectively in "the real world". The ever-present contrast between "the thing as we view it" and "the thing as it is" is the mechanism by which this crucially important distancing is accomplished. And maintaining this stance means that we are never entitled to claim to have exhausted a thing *au fond* in cognitive regards, to have managed to bring it wholly within our epistemic grasp. For to make this claim would, in effect, be to *identify* "the thing at issue" purely in terms of "our own conception of it", an identification which would effectively remove the former item (the thing itself) from the stage of consideration as an independent entity in its own right, by endowing our conception with decisively determinative force. And this would lead straightaway to the unacceptable result of a cognitive solipsism that would preclude reference to inter-subjectively identifiable particulars, and would thus block the possibility of inter-personal communication and communal inquiry.

Any pretentions to the predominance, let alone the correctness, of our own conceptions regarding the realm of the real must be set aside in the context of communication. In communication regarding things we must be able to exchange information about them with our contemporaries and to transmit information about them to our successors. And we must be in a position to do this on the presumption that *their* conceptions of things are not only radically different from *ours*, but conceivably also rightly different. Thus, it is a crucial precondition of the possibility of successful communication about things that we must avoid laying any claim either to the completeness or even to the ultimate correctness of our own conceptions of any of the things at issue. This renders it critically important *that* (and understandable *why*) conceptions are not pivotal for communicative purposes. Our discourse *reflects* our conceptions and perhaps *conveys* them, but it is not substantively *about* them.

What is crucial for communication, however, is the fundamental intention to deal with the objective order of this "real world". If our assertoric commitments did not transcend the information we have on hand, we would never be able to "get in touch" with others about a shared objective world. No claim is made for the *primacy* of our conceptions, or for the *correctness* of our conceptions, or even for the mere *agreement* of our conceptions with those of others. The fundamental intention to discuss "the thing itself" predominates and overrides any mere dealing with the thing as we conceive it to be. Certainly, that reference to "objectively real things" at work in our discourse does not contemplate a peculiar sort of *thing* – a new *ontological* category of "things-in-themselves". It is simply a shorthand formula for a certain communicative presumption or imputation rooted in an *a priori* commitment to the idea of a commonality of objective focus that is allowed to stand unless and until circumstances arise to render this untenable.

How do we really know that Anaximander was talking about *our* earth? He is not here to reassure us. He did not leave elaborate discussions about his aims and purposes. How can we be so confident about what he meant in that strange talk about a slab-like object suspended in equilibrium in the center of the cosmos? The answer is straightforward. That he is *to be taken* to mean that *our* earth is such an object is something that turns, in the final analysis, on two very general issues in which Anaximander himself plays little if any role: (1) our subscription to certain generalized principles of interpretation with respect to the Greek language; and (2) the conventionalized subscription by us and ascription to other language-users in general of certain fundamental communicative policies and intentions. In the face of appropriate functional equivalences we allow neither a difference in language nor a difference of "thought-worlds" to block an identity of reference.[4]

Seen in *this* light, the key point may be put as follows: it is indeed a presupposition of effective communicative discourse about a thing that we purport (claim and intend) to make true statements about it. But for such discourse it is *not* required that we purport to have a true or even adequate conception of the thing at issue. On the contrary, we must deliberately abstain from any claim that our own conception is definitive if we are to engage successfully in discourse. We deliberately put the whole matter of conceptions aside – abstracting from the question of the agreement of my conception with yours, and all the more from the issue of which one of us has the right conception.[5] This sort of epistemic humility is the price we pay for keeping the channels of communication open.

4. THE UTILITARIAN IMPERATIVE

Consider the following objection:

Let it be granted that this general approach is right – that the idea of a mind-independent reality is a presupposition basic to the conceptual framework that undergirds our project of inquiry and "knowledge" acquisition and communication. But why should one see this assumption as *validated* by its serviceability in this regard? After all, perhaps the entire project is simply unjustified. Consider the analogy of religion. God is essential to the project of religion and worship: the "external world" is essential to the project of inquiry and cognition. But perhaps those entire projects are simply inappropriate.

In countering this considered objection with respect to cognition, we must stress the inappropriateness of the analogy. For the religious project is optional, one may simply decline to enter in. But the cognitive project is not so easily evaded. We must act to live: must eat this or that, move here or there, do something or other. And, being the sort of creatures we are, our actions are guided by our beliefs. Is this substance edible? Is that place safe? Is that action goal-conducive? If we do not form views on these subjects and allow our actions to be guided by "knowledge" – or pretentions thereto – then there are but few alternatives (all duly noted and recommended by the sceptics of classical antiquity):

– to follow custom and "do what is generally done";
– to follow instinct or "hunch";
– to follow our desires and the modes of "inclination";
– to be guided by "probabilities";

But none of these non-cognitive alternatives seem very promising, none have much appeal to a creature who demands good reasons for acting. (Even to begin to validate a reliance on probabilities we need facts.)

The impetus to inquiry for knowledge-acquisition reflects the most practical of imperatives. Our need for *intellectual* accommodation in this world is no less pressing and no less *practical* than our need for physical accommodation. But in both cases, we do not want just some house or other, but one that is well built, that will not be blown down by the first wind that sweeps along. Sceptics from antiquity onward have always said, "Forget about those abstruse theoretical issues; focus on your practical needs." They overlook the crucial fact that an intellectual accommodation to the world is itself one of our deepest practical needs – that in a position of ignorance or cognitive dissonance we cannot function satisfactorily.

The project of communal inquiry is not optional – at any rate not for us humans. Its rationale lies in the most practical and prudent of considerations, since it is only by travelling the path of inquiry that we can arrive at the sorts of good reasons capable of meeting the demands of a "*rational* animal". And given the mandatory nature of the cognitive project, we have no real choice but to "buy in" on its presuppositions.

We thus arrive at an overall course of justificatory argumentation whose structure runs as follows:

(1)   We cannot survive and flourish in this world without effective action.
(2)   We cannot act effectively without rationally warranted confidence in our (putative) knowledge.
(3)   We cannot achieve confidence-inspiring knowledge without rational inquiry.
(4)   Commitment to a real world is an essential requisite for rational inquiry.

Therefore:

(5)   Realism (i.e. commitment to a real world that is the object of our inquiries) is a rational imperative on the side of practical reason – a *sine qua non* for a rational creature like ourselves to survive and to flourish.

Only in subscribing to that fundamental reality postulate can we take the sort of view of experience, inquiry, and communication that we in fact have. Without it, the entire conceptual framework of our thinking about the world and our place within it would come crashing down. The utility of the conception of reality is such that even if reality were not there, we would have to invent it.

5. RETROJUSTIFICATION: THE WISDOM OF HINDSIGHT

How can functional utility by itself provide an adequate validation? A "validation" in terms of functional utility establishes our claims to mind-independent reality not by the cognitive route of learning but by the pragmatic route of an eminently useful postulation. Clearly this cannot be the *entire* story.

The consideration that we *must* proceed in the way of objectivity-presuming cognition as a matter of functional requiredness, seeing that there is just no alternative if our aims are to be attained and our purposes served, stops well short of being totally satisfactory. It does not offer us any assurance that we actually will succeed in our endeavor if we do proceed in this way; it just has it that we will not if we do not. The issue of actual effectiveness remains untouched. But a nagging doubt still remains. It roots in the following challenge:

Let us grant that this line of approach provides a cogent practical argument. All this shows is that realism is *useful*. But does that make it *true*? Is there any rational warrant for it over and above the mere fact of its utility?

At this point we have to move beyond presupposed functional requisites to address the issue of actual effectiveness. We must now have recourse to the resources of actual experience. For what *is* learned by experience – and can only be learned in this way – is that in proceeding on this prejudgment our attempts do, by and large, work out pretty well *vis-à-vis* the purposes we have in view for inquiry and communication. When it comes to this issue of actual efficacy, we have no choice but to proceed experientially – through the simple strategem of "trying and seeing". Functional requiredness remains a matter of *a priori* considerations, but efficacy – actual sufficiency to our purposes – will be a matter of *a posteriori* experience. It is, and is bound to be, a matter of retrojustification – a retrospective revalidation in the light of experience. And this empirically delivered pragmatic consideration that our praxis of inquiry and communication does actually work – that we can effectively and (by and large) successfully communicate with one another about a shared world, inquiry into whose nature and workings proceeds successfully as a communal project of investigation – is the ultimately crucial consideration that legitimates (through "retrovalidation") the evidence-transcending imputations built into the objective claims to which we subscribe.

We want and need objective information about "the real world". This, of course, is not to be had directly without the epistemic mediation of experience. And so we treat certain data as evidence – we extend "evidential credit" to them as it were. Through trial and error we learn that some of them do indeed *deserve* it, and then we proceed to extend to them greater weight – we "increase their credit limit" as it were and rely on them more extensively. And, of course, to use those data as evidence is to build up a picture of the world, a picture which shows, with the "wisdom of hindsight", how appropriate it was for us to use those evidential data in the first place.

Charles Sanders Peirce put the problem with characteristic clarity: "It may be asked how I know that there are reals. If this hypothesis is the sole support of my method of inquiry, my method of inquiry must not be used to support my hypothesis."[5] Peirce put his finger on exactly the right question. Yet while this reality-hypothesis is indeed not a product of inquiry, but a presupposition for it, nevertheless, it is one whose justification ultimately stands or falls on the success of the inquiries it facilitates. Its validation cannot be preestablished through evidence but can only be provided *ex post facto* through the justificatory impetus of successful implementation.

What we began with was a basic project-facilitating postulation. But this does not tell the whole of the justificatory story. For there is also the no less important fact that this postulation obtains a vindicating retro-justification because the farther we proceed on this basis, the more its obvious appropriateness comes to light. With the wisdom of hindsight we come to see with increasing clarity that the project that these presuppositions render possible is an eminently successful one. The pragmatic turn does crucially important work here in putting at our disposal a style of justificatory argumentation that manages to be cyclical without vitiating circularity.

Accordingly, the substantive picture of nature's ways that is secured through our empirical inquiries is itself ultimately justified, retrospectively as it were, through affording us with the presuppositions on whose basis inquiry proceeds. As we develop science there must come a "closing of the circle". The world-picture that science delivers into our hands must eventually become such as to explain how it is that creatures such as ourselves, emplaced in the world as we are, investigating it by the processes we actually use, should do fairly well at developing a workable view of that world. The "validation of scientific method" must in the end itself become scientifically validated. Science must (and can) retrovalidate itself by providing the materials (in terms of a science-based world-view) for justifying the methods of science. (How this is to be done has already been sketched out in Chapter Seven.) Though the process is cyclic and circular, there is nothing vicious and vitiating about it.

The rational structure of the overall process of justification looks as follows.

(1) We use various sorts of experiential data as evidence for objective fact.

(2) We do this in the first instance for *practical* reasons, *faute de mieux*, because only by proceeding in this way can we hope to resolve our questions with any degree of rational satisfaction.

(3) As we proceed two things happen:

    (i) On the pragmatic side we find that we obtain a world picture on whose basis we can operate effectively. (Pragmatic revalidation.)

    (ii) On the cognitive side we find that we arrive at a picture of the world that provides an explanation of how it is that we are encouraged to get things (roughly) right – that we are in fact justified in using our phenomenal data as data of objective fact. (Explanatory revalidation.)

Accordingly, the success at issue is twofold – both in terms of understanding (cognition) and in terms of application (praxis). And it is this ultimate success that justifies and rationalizes, retrospectively, our evidential proceedings.

We arrive at the overall situation of dual "retrojustification" given in Figure 1, which shows that the presuppositions of inquiry are ultimately justified because a "wisdom of hindsight" enables us to see that by their means we have been able to achieve both practical success and a theoretical understanding of our place in the world's scheme of things. This includes how our inquiry methods manage to succeed. The cycles must – and presumably do – close in smooth loops of systemic justification. And both loops are crucial. Successful practical implementation is needed as an extra-theoretical quality-control monitor of our theorizing. And the capacity of our scientifically devised view of the world to underwrite an explanation of how it is that a creature constituted as we are, operating by the means of inquiry that we employ, and operating within an environment such as ours, can ultimately devise a relatively accurate view of the world is also critical for the validation of our knowledge.[7] The closing of these inquiry-geared loops validates, retrospectively, those realistic presuppositions or postulations that made the whole process of inquiry possible in the first place. Realism thus emerges as a presupposition-affording postulate for inquiry – a postulation whose ultimate legitimation eventuates retrospectively through the results, both practical and cognitive, which the process of inquiry based on those yet-to-be-justified presuppositions is able to achieve.

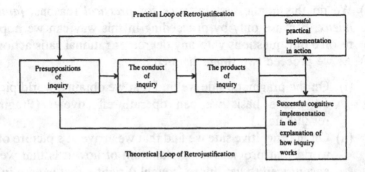

Note: Such a cycle explains, with the hindsight afforded by the products of inquiry, how successful inquiry is possible.

Fig. 1. The retrojustification of the presuppositions of inquiry

Let us review the overall line of deliberation. Metaphysical realism – the doctrine that there is a mind-independent reality and that our experience provides us with a firm cognitive grip upon it – does not represent a learned fact but a presuppositional postulate. As such, it has a complex justification that comes in two phases.

The first, *initial* phase is prospective, proceeding with a view to the functional necessity of *taking* this position – its purpose-dictated inevitability. For this step alone renders possible a whole range of activities relating to inquiry and to communication that is of the highest utility for us – and indeed is a practical necessity. In possibilizing[8] – that is, bringing it within the range of the feasible – a host of purpose-mandated activities, the postulate of metaphysical realism obtains its initial justification in the practical order of reasoning.

Such initial functional justification is good but not good enough. The second phase of justification goes further, albeit retrospectively. It proceeds by noting that after we actually engage in the goal-directed practice that the postulate in question possibilizes, our applicative and explanatory efforts are, in fact, attended by success – that making the initial postulate has an immense pragmatic payoff. This issue of actual efficacy is ultimately crucial for the overall justification of the practical postulate at issue.

A notion that has such important work to do cannot be dismissed as vacuous or superfluous. As was observed above, the utility of the conception of reality is so great that if it were not already there we would have to invent it. But the pragmatic success that ensues when we put this conception to work goes to show that we have not in fact done so.[9]

CHAPTER ELEVEN

INTIMATIONS OF IDEALISM

SYNOPSIS. (1) Realism rests on the pragmatic idea that if we did not *take* our experience to serve as an indication of objective facts, then we just could not validate any objective claims whatsoever. Its justification rests initially on this strictly functional basis. (2) Of course, once we make this initial assumptive leap we can do much better. We can *retrospectively use the teachings of science* to account for how the acquisition of scientific information of the real is possible for us. But this canonization of the theory-claims of our science requires the idealized stance that our science is substantially correct. (3) Accordingly, the sort of realism we espouse is committed to a cognizability-in-principle standard of real/true/actual that takes the lie of a *conceptual* idealism which holds that existence is to be explicated in mind-referring terms of reference. (4) But man is not the measure. We cannot justifiably hold that *only* what we can know can possibly qualify as real. Human reality is not necessarily the only sort there is. (5) The crucial point emerges that the realism/idealism debate is rendered complex through the great variety of realisms and opposing idealisms that can be contemplated. The controversy hinges on complex distinctions, it is not a straightforward choice of either/or. The overall picture is a complex one that leaves a great deal to be said on idealism's behalf.

1. THE IDEALISTIC ASPECT OF METAPHYSICAL REALISM

The ontological thesis that there is a mind-independent physical reality to which our inquiries address themselves more or less adequately is the key contention of realism. The preceding deliberations have argued that this basic thesis has the epistemic status of a presuppositional postulate that is initially validated by its pragmatic utility and ultimately retrovalidated through the results of its implementation (in both practical and theoretical respects). Our commitment to realism is, on this account, initially not based on *inquiries* about the world, but reflects a facet of how we *conceive* the world. The sort of realism contemplated here is one that pivots on the fact that we think of reals in a certain sort of way – that to say something is real is to credit it with a mind-independent status.

A position of this sort is in business as a realism all right. But seeing that it pivots on the character of our concepts and their *modus operandi*, it transpires that the business premises it occupies are to some extent

146

mortgaged to idealism. The fact that a meaningful realism has to be *a realism of things as we conceive them to be* allows idealism to infiltrate into the realist's domain.

Initially, at any rate, realism is an *input* into our investigation of nature rather than an output thereof. It does not represent a discovered fact, but a methodological presupposition of our praxis of inquiry; it is not constitutive (fact-descriptive) but regulative (praxis-facilitating). In the first instance, realism is a position to which we are constrained not by the push of evidence but by the pull of purpose. Here proceeding not with reference to the consequences of prior evidence but with reference to the preconditions of our objectives, its direction of movement is not from evidence but towards goals. It is not a factual discovery, but a practical postulate justified by its utility or serviceability in the context of our aims and purposes, seeing that if we did not *take* our experience to serve as an indication of facts about an objective order we would not be able to validate any objective claims whatsoever.[1]

Now insofar as realism stands on this pragmatic basis, it does not rest on considerations of independent substantiating evidence about how things actually stand in the world, but rather it is established by considering, as a matter of practical reasoning, how we do (and must) think about the world within the context of the projects to which we stand committed. Such a position sees this commitment to a mind-independent reality in an essentially utilitarian role – as a functional requisite for our intellectual resources (specifically for our conceptual scheme in relation to communication and inquiry). Thanks to its enmeshment in considerations of aims and purposes, it is clear that this sort of commitment to an objectivistic realism harks back to the salient contention of classical idealism that values and purposes play a pivotal role in our understanding of the nature of things. Seeing that a pragmatic line of approach pivots the issue on what is useful for us and productive for us in the context of our evaluatively legitimated aims and purposes, we return to the characteristic theme of idealism – the active role of the knower not only in the constituting but also in the constitution of what is known.

To be sure, this sort of idealism is not substantive but methodological. It is not a denial of real objects that exist independently of mind and as such are causally responsible for our objective experience. Quite the reverse, it is designed to facilitate their acceptance. But it insists that the justificatory *rationale* for this acceptance lies in a framework of mind-supplied purpose. For our mind-independent reality arises not *from* experience but *for* it – i.e. for the sake of our being in a position to exploit our experience to ground inquiry and communication with respect to the objectively real.

Accordingly, what we have here is an object-level realism that rests on a presuppositional idealism at the justificatory infralevel. We arrive, paradoxical as it may seem, at a realism that is founded – initially at least – on a fundamentally idealistic basis, a realism whose *justificatory basis* is ideal.

2. THE IDEALISTIC ASPECT OF EPISTEMOLOGICAL REALISM

The preceding deliberations pertain to the initial pragmatic justification of realism. The retroactive part of the validation of realism must also be considered. For, as we have seen, once scientific information is at hand, we can retrospectively *use* the teachings of science to account for how it is that the acquisition of scientific information is possible for us.

To be sure, one cannot move from scientific knowledge-claims to the objective characterization of reality without the mediating premiss that these claims are substantially correct. And this mediating premiss is not available with respect to existing science – science in the *present* state-of-the-art – but only with respect to *ideal* science. Only in the idealized case of an unrealistic perfection can we unproblematically adopt the stance of a theory-realism that holds that the world actually is as theorizing claims it to be. The cannonization of the theory-claims of science as reality-descriptive requires the idealized stance that science is substantially correct.

Strange though it may seem, this *ex post facto* realism takes the vantage point of idealization to make scientific realism a tenable proposition. And so here also, with the second, the epistemic component of realism we encounter a presupposition of an essentially idealistic sort. For it is only at the level of idealization that fully adequate knowledge of reality can be expected: only that idealization presented by perfected science manages "to tell it as it is". With respect to ideal science we can be realistic, but with respect to real science we must be idealists. The world-picture of current natural science need certainly not characterize reality and cannot validly be presumed to do so. Its world – the world it envisions *as it envisions it* – may well lie only in the minds of the beholders and must, in fact, be presumed to do so. In this regard too, then our realism has an idealistic mien.

3. CONCEPTUAL IDEALISM

The realist holds the existence and nature of reality to be mind-independent. But independent of *whose* mind? One can, in theory, hold that real things are mind-independent only in the sense of existing independently of,

(1)   *My* conceptions of them.
(2)   *Our* current (communal) conceptions.
(3)   *Anyone's* conception (ever).
(4)   *Their conceivability-in-principle.*

As we move down this list, the position we take becomes increasingly more problematic. With (1) we start with the rejection of megalomania. Well and good! But with (4) we reject Peirce's condemnation of incognizables, and this is something far more questionable. To accept the existence of things that cannot *in principle* be known – things which by their very nature lie outside the range of any and every (hypothetically possible) intelligence – is strong stuff.

Berkeley maintained that "to be (real) is to be perceived" (*esse est percipi*). This does not seem all that plausible. It seems more sensible to adopt "to be is to be perceivable" (*esse est percipile esse*). For Berkeley, of course, this was a distinction without a difference: if it is perceivable at all, then God perceives it. But if we embargo philosophical reliance on God, the matter looks different. We are then driven back to the question of what is perceivable for perceivers who are *physically realizable* in "the real world". And so, something really exists if it is, in principle, experientiable: "To be (physically) real is to be possible object of perception of a possible perceiver – one who is physically realizable in the world." *Physical* existence is seen as tantamount to observability-in-principle. The basic idea is that one can only claim (legitimately or appropriately) that a particular physical object exists if there is experiential access to it – not necessarily for us but for some experience-capable sort of creature.

On this approach, one would endorse the idea that to be part of physical reality is to be:

–   not necessarily observed, but observable;
–   not necessarily perceived, but perceivable;
–   not necessarily experienced, but experientiable.[2]

If something indeed exists in the world, then it must be observable-in-principle, detectable by some suitably endowed creature equipped with some suitably powerful technology. To exist (physically) is to be part of the world's causal commerce – to be at the initiating or receiving end of a causal process that can, in principle, be detected and monitored by an attentive intelligence. "Observability" (as contradistinguished from actual observation) is clearly something *objective* – a matter of what beings with mind-endowed capacities *can* do (can experientially manage), and not with what any particular one or more of them actually *does* experience. Accordingly, reality or existence does not involve a physical dependency on actual minds, but rather a *conceptual* dependency on the possibility of mental access. The very notion of observability (perceivability, experientiability) is *constituted* in a way that makes reference to a mind-operated process. And so, to take this position and to coordinate real existence with a mental process (i.e. observation, perception, experimental encounter, or the like) is to become a "*conceptual* idealist" in the sense of according mind-involving operations a key role in the explanation of what it is to be real.

Compare and contrast the following three theses:

(1)   To be real is to be recognized as such by a (real) mind – i.e. to exist *for* an actual mind.

(2)   To be real is to be recognizable as such by a (possible) mind – to be accessible to a (possible) mind.

(3)   To be known to be real is to be known to be such by a mind.

The first of these theses represents an *ontological* idealism ("to exist is to exist for a mind"). The last thesis is a mere trivial truism (only mind endowed beings can, by hypothesis, know anything). But the second represents a distinct intermediate position – a *conceptual* idealism which holds that the appropriate explanation of what it is for something to exist physically must be given in terms that make reference to mind.[3]

Bertrand Russell said that "idealists tell us that what appears as matter is really something mental."[4] But that is rather stretching the matter. (It is akin to validating "A dog is an animal that barks loudly" by allowing only conveniently screened individuals to belong to the type.) Idealism need not go so far as to say that mind *makes or constitutes* matter. It is quite enough to say that the salient and characteristic properties of physical existents are like sensory properties in that they represent dispositions to affect mind-endowed creatures in a certain sort of way, so that these properties have no standing at all in contexts where minds cannot be brought into the range of discussion.

In the light of this analysis, the present discussion has a mixed or mediative character. It is a realism all right, since it acknowledges a realm of ontological mind-independent existence. But since it stands committed to a cognizability-in-principle standard of real/true/actual it is also a *conceptual* idealism which holds that existence is to be explicated in mind-referring terms of reference.

4. IS MAN THE MEASURE?

Whatever can be *known* by us humans to be real will, *ipso facto*, be real. But does the converse hold? Is it appropriate to join C. S. Peirce who, in rejecting "incognizables", insisted that whatever is real at all must in principle be knowable by us humans – that *only* what we can know can possibly qualify as real?

Is humanly cognizable reality the only sort of reality there is? Some philosophers certainly say so, even maintaining that there actually is a fact of the matter only when a claim to this effect is such that "we [humans] could in finite time bring ourselves into a position in which we were justified either in asserting or in denying [it]".[5] On such a view reality is *our* reality. What we humans are not in a position to domesticate cognitively – what cannot be brought home to us by (finite!) cognitive effort – simply does not exist as a part of reality at all. Where we have no cognitive access, there just is nothing to be accessed. We are led back to the *homo mensura* doctrine of Protagoras: "Man is the measure of all things, of what is, that it is, of what is not, that it is not."

However, in reflecting on the issue in a more modest mood, one is tempted to ask: Just who has appointed us to this exalted role? How is it that *we humans* are established as the ultimate arbiters of reality as such?

On this issue traditional realism takes an appropriately modest line. It insists on preserving, insofar as possible, a boundary-line of separation between ontology and epistemology; between fact and knowledge of fact with respect to states of affairs; between truth-status possession and truth-status decidability with respect to propositions; between entity and observability with respect to individual things. As the realist sees it, reality can safely be presumed to have depths that human cognition may well be unable to plumb. (Extreme realists, to be sure, not only reject Berkeley's "to be *is* to be perceived" but even hesitate over its – here endorsed! – far weaker cousin, "to be is to be perceivable [by someone or other]".)

We can attenuate, of course, the fact/cognition distinction by liberalizing cognizers. Consider the following series: *For something to be a fact it is necessary that it be knowable by,*

(1)   *Oneself.*
(2)   *One's contemporary (human) fellow inquirers.*
(3)   *Us humans (at large and in the long run).*
(4)   *Some actual species of intelligent creatures.*
(5)   *Some physically realizable (though not necessarily actual) type of intelligent being – creatures conceivably endowed with cognitive resources far beyond our feeble human powers.*
(6)   *An omniscient being (i.e. God).*

The "*i*-th level" idealist maintains, and the "*i*-th level" realist denies such a thesis at stage (*i*). On this approach, the idealist emerges as the exponent of a verifiability theory of reality, equating truth and reality with what is verifiable by "us" – with different, and increasingly liberal constructions of just who is to figure in that "us group".

Where one sets the boundary in interpreting the idealist doctrine that to be real is to be cognizable will determine the sort of idealism that is at issue. No *sensible* idealist maintains a position as strong as (1). No *sensible* realist denies a position as weak as (6). The salient question is just where to draw the line in determining what is a viable "realistic/idealistic" position.

Let us focus upon case (3), the "man is the measure," *homo mensura* doctrine. Of course, what we humans can *know* to be real constitutes (*ex hypothesi*) a part or aspect of reality-at-large. That much is not in question. The bone of contention between *homo mensura* realism and idealism is the question of a surplus – of whether reality may have parts or aspects that outrun the reach of human cognition. By *this* standard, both Peirce and the Dummett of the preceding quotation on page 151 are clearly *homo mensura* realists, seeing that both confine the real to what *people* can, in principle, know.

It seems sensible to take the stance that a naturally evolved mind has a sufficiently close link to reality as to be able to secure some knowledge of it. But the converse is eminently problematic. It is dubious that the linkage should be so close that only what is knowable for some actual being should be real – that reality has no reserve surplus of fact and conditions that is not domesticable within the cognitive range of existing creatures (let alone one particular species thereof!). Accordingly, it seems sensible to adopt the "idealistic" line only at level (5), and to be a realist short of that.

Our own position, in any case, is a stance geared to the situation at level (5). It takes the line that "to be real is to be causally active – to be a part of the world's causal commerce". And since one can always hypothesize a creature that detects a given sort of causal process, we have not hesitated to equate reality with experientiability in principle. We thus arrive at a level (5) idealism, one of the very weakest viable sorts. At all the lower-numbered, more substantive levels our position is effectively realistic. But, of course, this approach is a halfway-house position. Those who believe that a plausible line could be maintained after (2) but still short of (5) would class its correlative position as a *scientific idealism* rather than a "genuine" realism.

A qualified realism of this sort holds that what is so as a "matter of fact" is not necessarily cognizable by "us" no matter how far within the limits of plausibility we extend the boundaries of that "us-community" of inquiring intelligences, as long as we remain within the limits of the actual. Any particular sort of possible cognizing being can know only a part or aspect of reality. But the situation changes when we move from *any* to *all*. One cannot make plausible sense of "such-and-such a feature of nature is real but no possible sort of intelligent being could possibly discern it." To be real is to be in a position to make an impact somewhere on something of such a sort that a suitably equipped mind-endowed intelligent creature could detect it. What is real in the world must make some difference to it, that is *in principle* detectable. Existence-in-this-world is coordinated with perceivability-in-principle. And so, at this point, there is a concession to idealism.

In any case, *homo mensura* realism is untenable. There is no good reason to resort to a hubris that sees our human reality as the only one there is. Neither astronomically nor otherwise are we the center around which all things revolve. A *homo mensura* realism that equates the "that" and the "what" of real existence with what is knowable by us humans is totally implausible. After all, humans have the capacity not only for knowledge but also for imagination. And it is simply too easy for us to imagine a realm of things and states of things of which we can obtain no knowledge because "we have no way to get there from here", lacking the essential means for securing information in such a case.

5. CONCLUSION

Realism, as we have seen, embraces the following theses.

(1)   There is a mind-independent physical reality which, as such, has a descriptive nature of some sort.

(2)   We can know something about it – we can acquire (a substantial volume of) accurate information about the nature of the real.

(3)   This descriptive knowledge of reality characterizes it exactly as it is in itself – in terms of reference that are absolute (non-relative) and in no way hinge on some particular cognitive perspective, being independent of the particular ways and means used by inquirers in forming their picture of the real.

Owing to the internal complexity of the doctrine, there will be various stages to the realism/idealism controversy and very different versions of realism or idealism.

(i)   The issue of *metaphysical* realism/idealism hinges on the acceptance of (1) itself.

(ii)   The issue of *cognitive* realism/idealism hinges on whether one is prepared to go beyond (1) so far as to accept (2) as well.

(iii)   The issue of *descriptive* realism/idealism hinges on whether one is prepared to go beyond (2) to accept (3) as well.

The overall position that has been defended here is that realism is plausible through point (2) and ceases to be so at (3). It is at this more detailed "scientific" level that idealism comes into its own.

What is right about idealism is inherent in the fact that in investigating the real we are clearly constrained to use our own concepts to address our own issues – that we can only learn about the real in our own terms of reference. All that reality will ever provide us with are answers to the questions we put to it. But what is right about realism is that the answers to the questions we put to the real are provided by reality itself – whatever the answers may be, they are what they are because it is reality itself that determines them to be that way. Intelligence proposes but reality disposes.

The salient point to emerge from such deliberations is that the realism/idealism debate is rendered complex through the great variety of realisms and opposing idealisms that can be contemplated. When one considers the controversy in detached perspective, one is led to the recognition that there is no prospect of a one-sided victory here. The sensible move is to opt for the middle ground and to *combine* a plausible version of realism with a plausible version of idealism. The issue is not one of the dichotomous choice of *either* realism *or* idealism, but rather one of a compromise in the interests of a fruitful collaboration between these historically warring positions.

NOTES

NOTE TO PREFACE

[1] Recent discussions of scientific realism include: Wilfred Sellars, *Science, Perception, and Reality*, London, 1963; E. McKinnon (editor), *The Problem of Scientific Realism*, New York, 1972; Frederick Suppe (editor), *The Structure of Scientific Theories*, 2nd, ed., Urbana, 1977. For a general overview of the controversy which also provides extensive references to the diffuse literature of the topic see Jarrett Leplin (editor), *Scientific Realism*, University of California Press, Berkeley, Los Angeles, London, 1984.

NOTES TO CHAPTER ONE

[1] In the introduction to Hilary Putnam's *Philosophical Papers* (Vol. II, p. ix) Cambridge, 1980, there is a compact account of Richard Boyd's position according to which an empirical realism presupposes a principle to the effect that "terms in a mature science typically *refer*".

[2] The converse is, of course, possible. One can hold that the objects exist without subscribing to the idea that science provides a correct account of them. This plausible point is forcefully argued in Ian Hacking, *Representing and Intervening*, Cambridge, 1983, see esp. pp. 36–37.

[3] Ontological realism contrasts with ontological *idealism*; scientific realism contrasts with scientific instrumentalism: the doctrine that science in no way describes reality, but merely affords a useful organon of prediction and control.

[4] Karl R. Popper, *Objective Knowledge*, Oxford, 1972, p. 9.

NOTES TO CHAPTER TWO

[1] A homely fishing analogy of Eddington's is useful here. He saw the experimentalist as akin to a fisherman who trawls nature with the "net" of his equipment for detection and observation. Now suppose (says Eddington) that a real fisherman trawls the seas using a fishnet of two-inch mesh. Then fish of a smaller size will simply go uncaught. Similarly, the theorists who analyze the experimentalist's catch will have an incomplete and distorted view of aquatic life. Only by improving our observational means for "trawling" nature can such imperfections be mitigated.(See A. S. Eddington, *The Nature of the Physical World*, New York, 1928.)

[2] D. A. Bromley *et al.*, *Physics in Perspective*: Student Edition, NRC/NAS Publications, Washington D.C., 1973; p. 16.

[3] Gerald Holton, 'Models for Understanding the Growth and Excellence of Scientific Research', in Stephen R. Graubard and Gerald Holton (editors), *Excellence and Leadership in a Democracy*, New York, 1962, p. 115.

[4] Recall Goethe's stricture: "*Natur hat weder Kern noch Schale, Alles ist sie mit einem Male.*"

[5] In this regard, E. P. Wigner seems altogether correct in reminding us: "...that in order to understand a growing body of phenomena, it will be necessary to introduce deeper and deeper concepts into physics and that this development will not end by the discovery of the final and perfect concepts. I believe that this is true: we have no right to expect that our intellect can formulate perfect concepts for the full understanding of inanimate nature's phenomena." "The Limits of Science", *Proceedings of the American Philosophical Society*, Vol. 94, 1950, p. 424.

[6] The geographic exploration analogy is an old standby: "Science cannot keep on going so that we are always going to discover more and more new laws. . . . It is like the discovery of America – you only discover it once. The age in which we live is the age in which we are discovering the fundamental laws of nature, and that day will never come again". Richard Feynman, *The Character of Physical Law*, Cambridge, Mass., 1965, p. 172. See also Gunther Stent, *The Coming of the Golden Age*, Garden City, N.Y., 1969; and S. W. Hawking, 'Is the End in Sight for Theoretical Physics?' *Physics Bulletin*, Vol. 32, 1981, pp. 15–17.

[7] Marxist theoreticians take this view very literally – in the manner of Lenin's idea of the "inexhaustibility" of matter in *Materialism and Empirico-Criticism*. Purporting to inherit from Spinoza a thesis of the infinity of nature, they construe this to mean that any cosmology that denies the infinite spatial extension of the universe must be wrong.

[8] 'Remarks by D. Bohm', in *Observation and Interpretation*, edited by S. Körner, New York and London, 1957, p. 56. For a fuller development of Bohm's views on the "qualitative infinity of nature", see his *Causality and Chance in Modern Physics*, London and New York, 1957.

[9] "If by the 'infinite complexity of nature' is meant only the infinite multiplicity of the *phenomena* it contains, there is no bar to final success in theory making, since theories are not concerned with particulars as such. So too, if what is meant is only the infinite variety of natural phenomena . . . that too may be comprehended in a unitary theory," 'Scientific Revolutions for Ever?', *British Journal for the Philosophy of Science*, Vol. 19, 1967, p. 41. For a suggestive analysis of "the architecture of complexity", see Herbert A. Simon, *The Sciences of the Artificial*, Cambridge, Mass., 1969.

[10] The idea that our knowledge about the world reflects an *interactive* process, to which both the *object* of knowledge (the world) and the knowing *subject* (the inquiring mind) make essential and ultimately inseparable contributions, is elaborated in the author's *Conceptual Idealism*, Oxford, 1973.

[11] Compare D. A. Bromley's observation: "Even if physicists could be sure that they had identified all the particles that can exist, some obviously fundamental questions would remain. Why, for instance, does a certain universal ratio in atomic physics have the particular value 137.036 and not some other value? This is an experimental result: the precision of the experiments extends today to these six figures. Among other things, this number relates the extent or size of the electron to the size of the atom, and that in turn to the wavelength of light emitted. From astronomical observation it is known that this fundamental ratio has the same numerical value for atoms a billion years away in space and time. As yet there is no reason to doubt that other fundamental ratios, such as the ratio of the mass of the proton to that of the electron, are as uniform throughout the universe as is the geometrical ratio pi equals 3.14159. Could it be that such physical ratios are really, like pi, mathematical aspects of some underlying logical structure? If so, physicists are not much better off than people who must resort to wrapping a string around

a cylinder to determine the value of pi! For theoretical physics thus far sheds hardly a
glimmer of light on this question." D. A. Bromley *et al.*, p. 28.

[12] See the author's *Peirce's Philosophy of Science*, Notre Dame and London, 1978.

[13] The present critique of convergentism is thus very different from that of W. V. O.
Quine. He argues that the idea of "convergence to a limit" is defined for numbers but not
for theories, so that speaking of scientific change as issuing in a "convergence to a limit" is
a misleading metaphor. "There is a faulty use of a mathematical analogy in speaking of a
limit of theories, since the notion of a limit depends on that of a 'nearer than,' which is
defined for numbers and not for theories" *Word and Object*, New York, 1960, p. 23. I am
perfectly willing to apply the metaphor of substantial and insignificant differences to
theories, but am concerned to deny that, as a matter of fact, the course of scientific
theory-innovation must eventually descend to the level of trivialities.

[14] The ideas of this section are developed at greater length in the author's *Scientific
Progress*, Oxford, 1978.

NOTES TO CHAPTER THREE

[1] This, in effect, is the salient insight of twentieth century philosophy of science from C. S.
Peirce through K. R. Popper's *Logik der Forschung*, Vienna, 1935, to Nancy Cartwright's
How the Laws of Physics Lie, Oxford, 1984.

[2] Larry Laudan, 'The Philosophy of Progress', mimeographed preprint, Pittsburgh, 1979,
p. 4. *Cf. idem, Progress and its Problems*, Berkeley and Los Angeles, 1977.

[3] *Cf.* the author's *Peirce's Philosophy of Science*, Notre Dame and London, 1978.

[4] Peirce verges on seeing this point: but his latter- day congeners usually do not, and try to
get by with wholly transcendental arguments from the possibility of science. *Cf.* Wilfrid
Sellars, *Science and Metaphysics: Variations on Kantian Themes*, London, 1968.

[5] To say that some ideals *can* be legitimated by practical considerations is not to say that
all ideals *must* be legitimated in this way. On ideals and their ramifications see the author's
Ethical Idealism, Berkeley and Los Angeles, 1986.

NOTES TO CHAPTER FOUR

[1] Instrumentalists purport to be driven to this position on grounds of a commitment to
empiricism. But it is a strange sort of empiricism they espouse. Traditionally, empiricism
is the doctrine that all descriptive knowledge of the world must be grounded in experience. A doctrine which says that experience is impotent to provide for descriptive
knowledge of the real (extraphenomenal) world is surely an anti-empiricist doctrine, not
an empiricist one.

[2] *Cf.* Bas van Fraassen, *The Scientific Image*, Oxford, 1980, pp. 46, 68–9, and *passim*.

[3] We must even give up altogether on "physical objects". For physical object predicates
cannot be introduced into a phenomenalist framework without resorting to gap-filling
universals and counterfactuals in a way that is anathema to dedicated phenomenalists.
Compare J. J. C. Smart, *Philosophy and Scientific Realism*, London, 1983, pp. 22–25.

[4] Cicero, *De Finibus*, Book V, Chapter iv.

[5] Compare the author's *Methodological Pragmatism*, Oxford, 1977.

[6] Certainly, instrumentalists are often drawn to this position through positivist/empiricist sympathies, rather than fallibilist ones. But as the example of Hume already shows, the ultimate support of such a position itself generally lies in a scepticism based on historicist/relativist considerations.

[7] See Rudolf Carnap, 'The Aim of Inductive Logic' in Logic, Methodology, and Philosophy of Science, edited by E. Nagel, P. Suppes, and A. Torshin, Stanford, 1962, pp. 308–318.

[8] Karl Popper, Logik der Forschung, Vienna, 1935.

[9] Thomas Kuhn, The Structure of Scientific Revolutions University of Chicago Press, Chicago, 1970.

[10] This discussion draws on Chapter XII, "Scientific Realism," of the author's Empirical Inquiry, Totowa, 1982.

[11] Compare the criticism of Sellars given in footnote four to Chapter Three (see p. 158).

[12] Bas van Fraassen, op. cit, pp. 202–203.

[13] C. S. Peirce, Collected Papers, Vol. II, sect. 2.112. Compare William James: "a rule of thinking which would absolutely prevent me from acknowledging certain kinds of truth if those kinds of truth were really there, would be an irrational rule." The Will to Believe and Other Essays on Popular Philosophy, New York, 1986, pp. 27–28.

[14] Bas van Fraassen, op. cit., p. 24.

[15] See pp. 24–25 of the Introduction of Ernan McMullin (editor), Galileo: Man of Science, New York and London, 1967.

[16] Cited in Newton–Smith, The Rationality of Science, p. 29.

[17] Ernest Nagel, The Structure of Science, New York, 1961, p. 152.

NOTES TO CHAPTER FIVE

[1] Vagueness constitutes a context in which we trade off informativeness (precision) against probable correctness (security), with science moving towards the former and "everyday knowledge" towards the latter. The relevant issues are considered in tantalizing brevity in Charles S. Pierce's short discussion of the "logic of vagueness", which he laments as too much neglected.

[2] Hilary Putnam, 'What is Realism' in Jarrett Leplin (editor) Scientific Realism, Berkeley, Los Angeles, London, 1984.

[3] It is its far looser tie to theory that makes experimentation a bulwark of scientific realism. Compare Ian Hacking's Representing and Intervening, Cambridge, Mass., 1983, especially Chapter 16 "Experimentation and Scientific Realism".

NOTES TO CHAPTER SIX

[1] As one writer puts it, if they were not correct, then those scientific theories "could not be used in the explanation, prediction, construction, and diagnosis of the phenomena of ordinary life as they in fact are." Roy Bhaskar, A Realist Theory of Science, Atlantic Highlands, 1978, p. 64.

[2] Bertrand Russell, 'Dewey's New Logic', in P. Schilpp (editor), The Philosophy of John Dewey, New York, 1939, pp. 143–156.

[3] J. J. C. Smart, *Between Science and Philosophy*, Random House, New York, 1969, p. 150.

[4] This list is taken from Larry Laudan, 'A Confutation of Convergent Realism', *Philosophy of Science*, Vol. 48, 1981, pp. 19–49 (see p. 33). This essay, as well as its author's 'Explaining the Success of Science' in J. T. Cushing *et al.*, *Science and Reality*, Notre Dame, 1985, are highly relevant to the deliberations of the present chapter.

[5] William Newton Smith in M. Hollis and S. Lukes (editors), *Rationality and Relativism*, Cambridge, Mass., 1982, p. 119. Compare J. J. C. Smart. *loc. cit.*

[6] It is readily seen that this thesis is inter–deducible with the conjunction of its two predecessors.

[7] The *ad indefinitum* meliorism inherent in this thesis should be noted.

[8] The relative success of theories does not speak for their truth, but rather for the merits of the methods of inquiry by which they are substantiated. Compare the author's *Methodological Pragmatism*, Oxford, 1977, for this contrast between thesis pragmatism and method pragmatism.

NOTES TO CHAPTER SEVEN

[1] Obviously, no sensible relativism can maintain that "Everything is relative", without thereby self–destructing. Relativism must be developed with respect to a limited range; we cannot say "No proposition is to be asserted absolutely" but only "No proposition belonging to the range *R* is to be asserted absolutely" – where this proposition should, of course, not belong to *R*. And this is all we need for present purposes, seeing that the thesis "The contentions of natural science are somehow-relativized" is itself not a contention of natural science. It is a thesis about the domain, rather than one lying within it.

[2] Compare the discussion in Gosta Ehrensvard, *Man on Another World*, Chicago and London, 1965, pp. 146–148.

[3] His anthropological investigations pointed Benjamin Lee Whorf in much this same direction. He wrote: "The real question is: What do different languages do, not with artificially isolated objects, but with the flowing face of nature in its motion, color, and changing form; with clouds, beaches, and yonder flight of birds? For as goes our segmentation of the face of nature, so goes our physics of the cosmos." 'Language and Logic', in *Language, Thought, Reality*, edited by J. B. Carroll, Cambridge, Mass., 1956, pp. 240–241. Compare also the interesting discussion in Thomas Nagel, 'What is it Like to be a Bat?' in *Mortal Questions*, Cambridge, Mass., 1976.

[4] Thomas Kuhn, *The Structure of Scientific Revolutions*, Chicago, 1962.

[5] Georg Simmel, 'Uber eine Beziehung der Selektionslehre zur Erkenntnistheorie', *Archiv für systematische Philosophie und Soziologie*, Vol. 1, 1985, pp. 34–35 (see pp. 40–41).

[6] William James, *Pragmatism*, New York, 1907, p. 171.

[7] See E. Purcell in A. G. W. Cameron, editor, *Interstellar Communication: A Collection of Reprints and Original Contributions*, New York and Amsterdam, 1963.

[8] Paul Anderson, *Is There Life on Other Worlds?*, New York and London, 1963, p. 130.

[9] Christiaan Huygens, *Cosmotheoros: The Celestial Worlds Discovered – New Conjectures Concerning the Planetary Worlds, Their Inhabitants and Productions*, London, 1698, pp. 41–43.

[10] *Collected Papers*, 5.494.

[11] *Ibid*, 8.12.

[12] Christiaan Huygens, *op. cit.*, pp. 76–77.

[13] Roland Pucetti, *Persons: A Study of Possible Moral Agents in the Universe*, New York, 1969, *Cf*. Ernan McMullin in *Icarus*, 14, 1971, pp. 291–294.

[14] Loren Eiseley, *The Immense Journey*, New York, 1937.

[15] Fred Hoyle, *The Black Cloud*, New York, 1965.

[16] I. S. Shklovskii and Carl Sagan, *Intelligent Life in the Universe*, San Francisco, London, Amsterdam, 1966, p. 350.

[17] The preceding discussion draws upon Chapter 11 of the author's *The Limits of Science*, Berkeley and Los Angeles, 1985.

[18] Michael Dummett, *Truth and Other Enigmas*, Cambridge, Mass., 1978, p. 152.

NOTES TO CHAPTER EIGHT

[1] This very Kantian issue will be treated here in a very un-Kantian way. For the present deliberations will not be addressed, *à la* Kant, to certain *a priori* principles that supposedly *underlie* physics. Rather, our tale will unfold in terms of the factual (*a posteriori*) principles that *constitute* physics – the laws of nature themselves.

[2] Erwin Schroedinger, *What is Life?* Cambridge, 1945, p. 31.

[3] Eugene P. Wigner, 'The Unreasonable Effectiveness of Mathematics in the Natural Sciences,' *Communications on Pure and Applied Mathematics*, Vol. 13, 1960, pp. 1–4 (see p. 2).

[4] *Ibid.*, p. 14.

[5] Albert Einstein, *Lettres à Maurice Solovine*, Paris, 1956, pp. 114–115.

[6] K. R. Popper, *Objective Knowledge*, Oxford, 1972, p. 28.

[7] Mary Hesse, *Revolutions and Reconstructions in the Philosophy of Science*, Bloomington, 1980, p. 154.

[8] Conversations with Gerald Massey have helped in clarifying this part of the argument.

[9] G. Galilei, *Dialogo II, Le Opere di Galileo Galilei*, Edizione Nazionale, Vols. I–XX, Florence, 1890–1909, Vol. VII, p. 298. (I owe this reference to Jürgen Mittelstrass.)

[10] The preceding account draws upon the author's *The Riddle of Existence*, Lanham, 1984.

NOTES TO CHAPTER NINE

[1] Our position thus takes no issue with P. F. Strawson's precept that "facts are what statements (when true) state". ('Truth', *Proceedings of the Aristotelian Society*, Supplementary Vol. 24, 1950, pp. 129–156; see p. 136.) Difficulty would ensue only if an "only" were inserted.

[2] But can any sense be made of the idea of *merely* possible(i.e. possible but non-actual) languages? Of course it can! Once we have a generalized conception (or definition) of a certain kind of thing – be it a language or a caterpillar – then we are inevitably in a position to suppose the prospect of things meeting these conditions are over and above those that

in fact do so. The prospect of mooting certain "mere" possibilities cannot be denied – that, after all, is just what possibilities are all about.

[3] Note, however, that if a Davidsonian translation argument to the effect that "if it's sayable at all, then, it's sayable in *our* language" were to succeed – which it does not – then the matter would stand on a very different footing. For it would then follow that any possible language can state no more than what can be stated in our own (actual) language. And then the realm of facts (i.e. what is (correctly) statable in some *possible* language) and of that of truths (i.e. what is (correctly) statable in some *actual* language) would necessarily coincide. Accordingly, our thesis that the range of facts is larger than that of truths hinges crucially upon a failure of such a translation argument. (See Donald Davidson, 'The Very Idea of a Conceptual Scheme', *Proceedings and Addresses of the American Philosophical Association*, Vol. 47, 1973–1974, pp. 5–20, and also the critique of his position in Chapter II of the author's *Empirical Inquiry*, Totowa, 1982.)

[4] Compare Philip Hugly and Charles Sayward, 'Can a Language Have Indenumerably Many Expressions?' *History and Philosophy of Logic*, Vol. 4, 1983.

[5] One possible misunderstanding must be blocked at this point. To *learn* about nature, we must *interact* with it. And so, to *determine* a feature of an object, we may have to make some impact upon it that would perturb its otherwise obtaining condition. (The indeterminacy principle of quantum mechanics affords a well-known reminder of this.) It should be clear that this matter of physical interaction for data-acquisition is not contested in the ontological indifference thesis here at issue.

[6] For a useful survey of philosophical issues located in this general area, see Vincent Julian Fecher, *Error, Deception, and Incomplete Truth*, Rome, 1975.

[7] To be sure, *abstract* things, such as colors or numbers, will not have dispositional properties. For being divisible by four is not a *disposition* of sixteen. Plato got the matter right in Book VII of the *Republic*. In the realm of *abstracta*, such as those of mathematics, there are not genuine *processes* – and process is a requisite of dispositions. Of course, there may be dispositional truths in which numbers (or colors, etc.) figure that do not issue in any dispositional properties of these numbers (or colors, etc.) themselves – a truth, for example, such as my predilection for odd numbers. But if a truth (or supposed truth) does no more than to convey how someone *thinks* about a thing, then it does not indicate any property of the thing itself. In any case, however, the subsequent discussion will focus on *realia* in contrast to *fictionalia* and *concreta* in contrast to *abstracta*. (Fictional things, however, *can* have dispositions: Sherlock Holmes was addicted to cocaine, for example. Their difference from *realia* is dealt with below).

[8] This aspect of objectivity was justly stressed in the 'Second Analogy' of Kant's *Critique of Pure Reason*, though his discussion rests on ideas already contemplated by Leibniz, *Philosophische Schriften*, edited by C. I. Gerhardt, Vol. VII, pp. 319 ff.

[9] See C. I. Lewis, *An Analysis of Knowledge and Valuation*, La Salle, 1962, pp. 180–181.

[10] This also explains why the dispute over mathematical realism (Platonism) has little bearing on the issue of physical realism. Mathematical entities are akin to fictional entities in this – that we can only say about them what we can extract by deductive means from what we have explicitly put into their defining characterization. These abstract entities do not have non-generic properties since each is a "lowest species" unto itself.

[11] Compare F. H. Bradley's thesis: "Error *is* truth, it is partial truth, that is false only because partial and left incomplete." *Appearance and Reality*, Oxford, 1893, p. 169.

[12] The discussion of Sections 1–3 draws on Chapter 5 of the author's *Empirical Inquiry*, Totowa, 1982.

[13] C. S. Peirce, *Collected Papers*, 5.64–67. Compare 2.138 and:

Whenever I've come to know a fact, it is by its resisting us. A man may walk down Wall Street debating within himself the existence of an external world; but it in his brown study he jostles against somebody who angrily draws off and knocks him down. The sceptic is unlikely to carry his scepticism so far as to doubt whether anything besides the Ego was concerned in that phenomenon. The resistence shows that something independent of him is there. When anything strikes upon the senses the mind's train of thought is interrupted, for if it were not, nothing would distinguish the new observation from a fancy. (*Ibid.*, Vol. 1, sect. 1.431)

[14] *Meditations*, No. VI; *Philosophical Works*, edited by E. S. Haldane and G. R. T. Ross, Vol. I, Cambridge, 1911), pp. 187–189.

[15] See the author's *Scepticism* (Oxford, 1980).

NOTES TO CHAPTER TEN

[1] Kant held that we cannot experientially learn through perception about the objectivity of outer things, because we can only recognize our perceptions as perceptions (i.e. representations of outer things) if these outer things are taken as such from the first (rather than being learned or inferred). As he summarizes in the "Refutation of Idealism":

Idealism assumed that the only immediate experience is inner experience, and that from it we can only *infer* outer things – and this, moreover, only in an untrusworthy manner. . . . But in the above proof it has been shown that outer experience is really immediate . . . (CPuR, B276.)

[2] Maimonides, *The Guide of the Perplexed*, I, 71, 96a.

[3] The justification of such imputations is treated more fully in Chapter IX of the author's *Induction*, Oxford, 1980.

[4] Compare the interesting discussion in Michael E. Levin, 'On Theory-Change and Meaning-Change' in *Philosophy of Science*, Vol. 46, 1979, pp. 407–424.

[5] Charles S. Peirce, *Collected Papers*, 5.383.

[6] It is thus perfectly possible for two people to communicate effectively about something that is wholly non-existent and about which they have substantially discordant conceptions (for example, X's putative wife, where X is, in fact, unmarried, though one party is under the misimpression that X is married to A, and the other under the misimpression that X is married to B). The commonality of communicative focus is the basis on which alone the exchange of information (or misinformation) and the discovery of error becomes possible. And this inheres, not in the actual arrangements of the world, but in shared (conventionalized) intention to talk about the same thing.

[7] These issues are examined at some length in the author's *Methodological Pragmatism*, Oxford, 1977.

[8] In English, we have no single verb "to make possible" akin to the German *ermoeglichen*. To adopt "possibilize" would perhaps be sensible and certainly convenient.

[9] This chapter draws some relevant materials from the author's *Empirical Inquiry*, London, 1982.

NOTES TO CHAPTER ELEVEN

[1] In his 1932 paper 'Positivism and Realism', Moritz Schlick saw the claim that "something exists independently of us" as having an empirically testable meaning since the independent existence of atoms, cells, stars, historical events, etc., is to be seen as fully testable empirically because we can state the experiential conditions under which such hypotheses are confirmable. But the fact remains that if we did not precommit ourselves to taking empirical experience as indicative of objective facts, we could validate no factual claims whatsoever.

[2] Note that we cannot say "To be is to be describable" or "To be is to be identifiable", since purely hypothetical possibilities can also be described and identified. What they can not be is *experienced* by some physically realizable intelligence.

[3] For a fuller development of the implications of such a position, see the author's *Conceptual Idealism*, Oxford, 1973.

[4] Bertrand Russell, *The Problems of Philosophy*, Oxford, 1912, p. 58.

[5] Michael Dummett, 'Truth', *Proceedings of the Aristotelian Society*, Vol. 59 (1958–1959), p. 160.

BIBLIOGRAPHY

Roy Bhaskar, *A Realist Theory of Science* (Atlantic Highlands: Humanities Press, 1975).

Richard V. Boyd, 'Realism, Underdetermination, and a Causal Theory of Evidence,' *Nous*, 7 (1973), pp. 1–12.

Nancy Cartwright, *How the Laws of Physics Lie* (Oxford: Clarendon Press, 1983).

James T. Cushing *et al.*, *Science and Reality* (Notre Dame: University of Notre Dame Press, 1984).

Pierre Duhem, *The Aim and Structure of Physical Theory*, tr. by P. P. Wiener (Princeton: Princeton University Press, 1954).

Michael Dummett, *Truth and Other Enigmas* (Cambridge, MA: Harvard University Press, 1978).

A. S. Eddington, *The Nature of the Physical World* (London: Macmillan, 1928).

Herbert Feigl and Grover Maxwell (eds.), *Current Issues in the Philosophy of Science* (New York: Holt, Rinehart, and Winston, Inc., 1961).

James H. Fetzer, *Scientific Knowledge* (Dordrecht: D. Reidel, 1981).

Paul Feyerabend, *Against Method* (New York: Humanities Press, 1975).

Bas C. van Fraassen, 'The Charybdis of Realism: Epistemological Implications of Bell's Inequality,' *Synthese*, **52** (1982), pp. 25–38.

Bas C. van Fraassen, *The Scientific Image* (Oxford: Clarendon Press, 1980).

Ronald N. Giere, 'Constructive Realism' *in* P. M. Churchland and Clifford A. Hooker (eds.), *Images of Science* (Chicago: University of Chicago Press, 1985).

Clark Glymour, *Theory and Evidence* (Princeton: Princeton University Press, 1980).

Gary Gutting (ed.), *Paradigms and Revolutions* (Notre Dame: University of Notre Dame Press, 1980).

Ian Hacking, *Representing and Intervening* (Cambridge: Cambridge University Press, 1983).

R. J. Hirst (ed.), *Perception and the External World* (New York: Macmillan, 1965).

Clifford Hooker, 'Systematic Realism,' *Synthese*, **26** (1974), pp. 409–497.

William Kneale, 'Scientific Revolutions Forever?', *The British Journal for the Philosophy of Science*, **19** (1968), pp. 27–42.

Stephen Korner (ed.), *Observation and Interpretation* (New York: Academic Press, 1957).

Thomas Kuhn, *The Structure of Scientific Resolutions* (Chicago: University of Chicago Press, 1962).

Larry Laudan, *Progress and Its Problems* (Berkeley: University of California at Berkeley Press, 1977).

Jarrett Leplin (ed.), *Scientific Realism* (Berkeley, Los Angeles, London: University of California Press, 1984).

Michael E. Levin, 'On Theory-Change and Meaning-Change,' *Philosophy of Science*, **46** (1979), pp. 407–424.

Henry Margenau, *The Nature of Physical Reality* (New York: McGraw-Hill, 1950).

E. McKinnon (ed.), *The Problem of Scientific Realism* (New York: McGraw-Hill, 1972).

Ernest Nagel, *The Structure of Science* (New York: Harcourt Brace, 1961).

William Newton-Smith, *The Rationality of Science* (Boston: Routledge & Kegan Paul, 1981).

Karl Popper, *Conjectures and Refutations* (London: Routledge & Kegan Paul, 1963).

Karl Popper, *Logic of Scientific Discovery* (New York: Basic Books, 1959).

Karl Popper, *Objective Knowledge* (Oxford: Oxford University Press, 1972).

Hilary Putnam, *Mathematics, Matter, and Method, Philosophical Papers*, Vol. 1, (Cambridge: Cambridge University Press, 1975).

Hilary Putnam, *Mind, Language and Reality, Mathematical Papers*, Vol. 2, (Cambridge: Cambridge University Press, 1975).

W. V. O. Quine, *Word and Object* (Cambridge, Mass.: Technology Press of the Massachusetts Institute of Technology, 1960).

Hans Reichenbach, *Experience and Prediction* (Chicago: University of Chicago Press, 1938).

Hans Reichenbach, *Philosophic Foundations of Quantum Mechanics* (Berkeley and Los Angeles: University of California Press, 1944).

Nicholas Rescher, *Methodological Pragmatism* (Oxford: Blackwell, 1977).

Nicholas Rescher, *The Limits of Science* (Berkeley, Los Angeles, London: University of California Press, 1984).

Nicholas Rescher, *Conceptual Idealism* (Oxford: Blackwell, 1973).

Nicholas Rescher, *Empirical Inquiry* (Totawa: Rowman & Littlefield, 1982).

Nicholas Rescher, *Scientific Progress* (Oxford: Basil Blackwell, 1978).

Bertrand Russell, *Human Knowledge, Its Scope and Limits* (New York: Simon and Schuster, 1948).

Wesley C. Salmon, *Scientific Explanation and the Causal Structure of the World* (Princeton: Princeton University Press, 1984).

Israel Scheffler, *Science and Subjectivity* (New York: Bobbs-Merrill Co., 1967).

Wilfrid Sellars, *Science, Perception, and Reality* (London, New York: Humanities Press, 1963).

J. J. C. Smart, *Philosophy and Scientific Realism* (New York: Humanities Press, 1963).

Frederick Suppe (ed.), *The Structure of Scientific Theories*, 2nd ed. (Urbana: University of Illinois Press, 1974).

Patrick Suppes, 'What is a Scientific Theory?' *in* S. Morgenbesser (ed.), *Philosophy of Science Today* (New York: Basic Books, 1967).

Paul Teller, 'On Why-Questions,' *Nous*, **8** (1974), pp. 371–380.

Richard S. Westfall, *The Construction of Modern Science: Mechanisms and Mechanics* (New York: John Wiley and Sons, 1971).

INDEX OF NAMES

INDEX OF SUBJECTS